Economics and Politics of Energy

Economics and Politics of Energy

Edited by

Behram N. Kursunoglu
Global Foundation, Inc.
Coral Gables, Florida

Stephan L. Mintz
Florida International University
Miami, Florida

and

Arnold Perlmutter
University of Miami
Coral Gables, Florida

Springer Science+Business Media, LLC

Library of Congress Cataloging-in-Publication Data

Economics and politics of energy / edited by Behram N. Kursunoglu, Stephan L. Mintz and Arnold Perlmutter.
p. cm.
"Proceedings of the International Conference on Economics and Politics of Energy, held November 27-29, 1995, in Miami Beach, Florida."--T.p. verso.
Includes bibliographical references and index.

1. Power resources--Congresses. 2. Energy policy--Congresses. I. Kurşunoğlu, Behram, 1922- . II. Mintz, Stephan L. III. Perlmutter, Arnold, 1928- . IV. International Conference on Economics and Politics of Energy (1995 : Miami Beach, Fla.)
TJ163.15.E26 1996
333.79--dc20 96-27562
CIP

Proceedings of the International Conference on Economics and Politics of Energy, organized by Global Foundation, Inc., held November 27 – 29, 1995, in Miami Beach, Florida

ISBN 978-1-4757-8575-3 ISBN 978-0-585-34288-7 (eBook)
DOI 10.1007/978-0-585-34288-7

Originally published by Plenum Press, New York in 1996.
Softcover reprint of the hardcover 1st edition 1996

10 9 8 7 6 5 4 3 2 1

PREFACE

The 1995 conference was organized around two closely related themes and focused on the two pivotal aspects of energy, that is, economics and politics, both of which are decisive in providing long-term national and international strategies for the next century. Originally the program was going to include the participants from the new oil powers in Central Asia and Caucasus, newly independent from the former U.S.S.R. However, probably both economics and politics prevented their participation.

Global energy projections, technological changes such as nuclear power and the fuel geopolitics of the coming century will be the basis for political and strategic planning. Based on the scenarios of likely global economic and population growth and of new energy technologies, what are foreseeable scenarios for the geopolitics of energy a half century ahead? What fresh worldwide systems should we start now?

The political problems with profound economic impact could include, for example, the significance of the continuing worldwide growth of nuclear power, with such issues as the use of Highly Enriched Uranium (HEU) and Plutonium obtained from the dismantling of U.S. and former U.S.S.R. nuclear weapons; the urgency of nonproliferation; the disposal of civilian and military nuclear waste; and, nuclear power alternatives. In spite of U.S. reluctance, the increasing role of nuclear power is becoming apparent in several countries, and its potential has become an important political factor today.

The conference has made significant contributions to the above issues. These proceedings show some directions for the economics and politics of energy. Following this 18th in a series of energy conferences, the 1996 conference topics will include detailed discussions of nuclear energy: nuclear medicine, agricultural applications, reduction of the fossil fuel use of the past five decades resulting from the use of nuclear energy to generate electricity, economic impact of industrial applications, and analysis of the probable greenhouse gas accumulation during the next hundred years to ensue from burning all or most of the fossil fuel reserves of one trillion barrels of oil and the large reserves of coal in China, Russia, and the U.S.

Behram N. Kursunoglu
Stephen L. Mintz
Arnold Perlmutter
Coral Gables, Florida

ABOUT THE GLOBAL FOUNDATION, INC.

The Global Foundation, Inc., utilizes the world's most important resource... people. The Foundation consists of great senior men and women of science and learning and of outstanding achievers and entrepreneurs from industry, governments, and international organizations, along with promising and enthusiastic young people. These people form a unique and distinguished interdisciplinary entity, and the Foundation is dedicated to assembling all the resources necessary for them to work together. The distinguished senior component of the Foundation transmits its expertise and accumulated experience, knowledge, and wisdom to the younger membership on important global issues and frontier problems in science.

Our work, therefore is a common effort, employing the ideas of creative thinkers with a wide range of experience and viewpoints.

GLOBAL FOUNDATION BOARD OF TRUSTEES

GLOBAL FOUNDATION'S RECENT CONFERENCE PROCEEDINGS

Making the Market Right for the Efficient Use of Energy
Edited by: Behram N. Kursunoglu
Nova Science Publishers, Inc., New York, 1992

Unified Symmetry in the Small and in the Large
Edited by: Behram N. Kursunoglu and Arnold Perlmutter
Nova Science Publishers, Inc., New York, 1993

Unified Symmetry in the Small and in the Large - 1
Edited by: Behram N. Kursunoglu, Stephen Mintz, and Arnold Perlmutter
Plenum Press, 1994

Unified Symmetry in the Small and in the Large - 2
Edited by: Behram N. Kursunoglu, Stephen Mintz, and Arnold Perlmutter
Plenum Press, 1995

Global Energy Demand in Transition: The New Role of Electricity
Edited by: Behram N. Kursunoglu, Stephen Mintz, and Arnold Perlmutter
Plenum Press, 1996

Economics and Politics of Energy
Edited by: Behram N. Kursunoglu, Stephen Mintz, and Arnold Perlmutter

CONTRIBUTING CO-SPONSORS OF THE GLOBAL FOUNDATION CONFERENCES

Gas Research Institute, Washington, DC
General Electric Company, San Jose, California
Electric Power Research Institute, Palo Alto, California
Northrop Grumman Aerospace Company, Bethpage, New York
Martin Marietta Astronautics Group, Denver, Colorado
Black and Veatch Company, Kansas City, Missouri
Bechtel Power Corporation, Gaithersburg, Maryland
ABB Combustion Engineering, Windsor, Connecticut
BellSouth Corporation, Atlanta, Georgia
National Science Foundation
United States Department of Energy

International Conference
on
ECONOMICS AND POLITICS OF ENERGY
November 27-29, 1995
Doral Ocean Beach and Resort
REGENCY BALLROOM
Miami Beach, Florida

PROGRAM

MONDAY, November 27, 1995

8:30 AM **SESSION I:** **GLOBAL ENERGY DEMAND PROJECTIONS IN THE COMING CENTURY, DRIVEN BY POPULATION, ECONOMICS AND ENERGY EFFICIENCY**

Moderator: **BEHRAM N. KURSUNOGLU**, Global Foundation, Inc.
"Equipartition of the Nuclear Fuel Cycle as an Alternative for Non-Proliferation"

Keynote Dissertator: **RICHARD BALZHISER**, Electric Power Research Institute
"Global Energy and Electricity Futures"

Scientific Keynote Address: **EDWARD TELLER**, Lawrence Livermore Lab, UC
"Misconceptions Connected with Energy Policies"

Keynote Address: **GLENN T. SEABORG,** Lawrence Berkeley National Laboratory, California
"The Next Fifty Years of the Peaceful Applications of Nuclear Energy"

Annotator: **JOHN R. IRELAND,** Los Alamos National Laboratory, New Mexico
J. JEFFREY IRONS, Northrop Grumman Aerospace

Session Organizer: **BEHRAM N. KURSUNOGLU**

10:00 AM Coffee Break

10:15 AM **SESSION II:** **ENERGY SUPPLY PROJECTIONS OF PRIMARY FUELS, RENEWABLES, CONSERVATION, AND NUCLEAR POWER**

Moderator: **GERALD CLARK,** Uranium Institute, London

Dissertators: **STEVE FREEDMAN,** Gas Research Institute, Chicago
"The Role of Natural Gas in Electric Power Generation in the Twenty-First Century"
RAYMOND SERO, Westinghouse, Pittsburgh
"Potential Nuclear Energy Growth in Eastern Europe and Southeast Asia"
RANDY HUDSON, Oak Ridge National Laboratory, Oak Ridge, TN
"World Energy Use - Trends of Demand"

Annotators: **LEE ELDER,** Black and Veatch

Session Organizer: **WILLIAM M. JACOBI,** Monroeville, PA

MONDAY, November 27, 1995

11:15 AM **SESSION III:** **ELECTRICITY GENERATION GROWTH IN FREE MARKET DEVELOPING COUNTRIES, WITH PARTICULAR ATTENTION TO CENTRAL VS DECENTRAL SOURCES, RENEWABLES, AND COAL AND NUCLEAR**

Moderator: **M. JOHN ROBINSON,** Black and Veatch

Dissertators: **GERALD CLARK,** The Uranium Institute, London
"Nuclear Prospects in Southeast Asia and the Far East"
PHILIP W. COVELL, RICHARD D. HANSEN, AND JOSE G. MARTÍN, University of Massachusetts Lowell
"The Role of Subsidies and Private Investments in Sustainable Rural Electrification"
KIM HEINZ, Southern Electric International
"Globalization of the Electricity Industry"

Annotators: **JULIAN SANCHEZ GUTIERREZ,** National Institute of Nuclear Investigation, Mexico
MARIANO BAUER, Mexico

Session Organizer: **CAULTON L. IRWIN,** West Virginia University

1:00 PM Lunch Break

2:00 PM: **SESSION IV:** **THE NEXT FIFTY YEARS OF THE PEACEFUL APPLICATIONS OF NUCLEAR ENERGY**

Moderator: **BERTRAM WOLFE,** Monte Sereno, California

Dissertators: **LINDEN BLUE,** General Atomics Corporation, San Diego, California
"The Modular Helium Reactor and the Next Fifty Years of Nuclear Power"
SHELBY BREWER, ABB Combustion Engineering, Inc., CT
"Nuclear Power in Its Second Half Century"

POONG EIL JUHN, International Atomic Energy Agency, Vienna
"The Worldwide Perspectives for Nuclear Energy"
JOHN W. LANDIS, Stone and Webster International, Boston
"Medical Applications of Radio Isotopes and Nuclear Radiations"
TERRY R. LASH, U.S. Department of Energy
"Department of Energy Perspectives on the Future of Nuclear Power"
DON W. MILLER, Ohio State University
The INCS "Vision of the Second Fifty Years of Nuclear Science and Technology"

Annotators: **MARCELLO ALONSO,** Florida Institute of Technology, Melbourne, FL
JOHN W. ZONDLO, West Virginia University
JOHN IRELAND, Los Alamos National Laboratory

Session Organizer: **EDWARD ARTHUR**, Los Alamos National Laboratory

3:00 PM Coffee Break

MONDAY, November 27, 1995

5:00 PM **SESSION V:** **NUCLEAR POWER GROWTH, NON-PROLIFERATION, AND POLITICS OF NUCLEAR ENERGY**

Moderator: **RICHARD KENNEDY,** Washington, D.C.

Dissertators: **HAROLD BENGELSDORF,** Washington, DC
"The Special 'Blue Ribbon' Panel Report Issued by the American Nuclear Society on the Protection and Management of Plutonium"
STEVEN A. HUCIK, GE Nuclear Energy, San Jose, CA
"Advanced Boiling Water Reactor (ABWR)/First-of-a-Kind Engineering (FOAKE) Program"
JOSEPH D. LEHMAN, Lockheed Martin Astronotics, Denver
"Intelligence Requirements and Non-Proliferation"

J. ARTHUR de MONTALEMBERT, COGEMA, France
"The Civilian Recycling Industry: A Key Contribution to Disarmament"
JAMES TAPE, Los Alamos National Laboratory
"Nuclear Materials Safeguards for the Future"
PHILLIP SEWELL, U.S. Enrichment Corporation, Bethesda, Maryland
"U.S.-Russian HEU Agreement"

Annotators: **R. A. KRAKOWSKI,** Los Alamos National Laboratory, Los Alamos
JEAN-CLAUDE GUAIS, NUSYS, Paris

Session Organizer: **EDWARD ARTHUR,** LANL

7:00 PM Conference Adjourns for the day.

TUESDAY, November 28, 1995

8:30 AM **SESSION VI:** **INTERNATIONAL MANAGEMENT OF NUCLEAR POWER FUEL SYSTEMS**

Moderator: **HAROLD BENGELSDORF,** Washington, DC

Dissertators: **GORDON MICHAELS,** ORNL, TN
"Thermal Issues with the U.S. Geological Repository and the Potential Role for Waste Transmutation"
RICHARD WAGNER, JR., KAMAN Sciences Corp., Washington, D.C.
"Global Plutonium"
WILHELM GMELIN, Euratom, Belgium, Brussells
"The Effective and Future Roles of Nuclear Materials Controls in the Perspective of Developing a Society Friendly Nuclear Energy"
PIERRE ZALESKI, Université Paris, Dauphine
"Management of Russian Military or Weapons-Grade Plutonium"

Annotator: **JOSÉ MARTÍN,** University of Massachusetts, Lowell, MA

Session Organizer: **ANTHONY J. FAVALE**, Northrop Grumman Aerospace Co.

10:00 AM Coffee Break

10:15 AM Session Continues

12:00 PM: Conference Adjourns for the day. Lunch for the Trustees and Advisory Board Members of the Global Foundation in the Madrid Room, Mezzanine Level.

7:30 PM: Conference Banquet, Grand Promenade Room

WEDNESDAY, November 29, 1995

8:30 AM: **SESSION VII:** **RELEVANCE OF INTERNATIONAL CONSENSUS POLICIES ON ALTERNATIVE NATIONAL ENERGY STRATEGIES AS RELATED TO FREE MARKET ECONOMIC, GLOBAL FUEL TRANSPORTATION SYSTEMS FOR OIL AND GAS, ENVIRONMENTAL IMPACT AND GEOPOLITICAL TRENDS**

Moderator: **JEAN COUTURE,** Paris

Dissertators: **RAFET AKGÜNAY,** Turkish Embasssy, Washington, DC
"Transporting Oil from the Caspian Sea to Western Markets: A Turkish Perspective"
DAN HARTLEY, Sandia National Laboratory
"Renewables: A Key Component of Our Global Energy Future"
CURT MILEIKOWSKY, Stockholm
"On Global Warming, Future Living Standard and Environment"
RICHARD WILSON, Harvard University
"The Crucial Environmental Issues"

Annotators: **LAWRENCE F. DRBAL,** Black and Veatch
THOMAS CUNNINGTON, Cyton Corporation, Michigan

Session Organizer:

10:45 AM Coffee Break

11:00 AM <u>**SESSION VIII:**</u> **ROUND TABLE ON PROBABLE WORLDWIDE ENERGY SYSTEMS**

Round Table Moderator: **HENRY KING STANFORD,** President Emeritus of the Universities of Miami and Georgia

Round Table Dissertators: **SHELBY BREWER, GERALD CLARK, STEVE FREEDMAN, POONG EIL JUHN, RICHARD KENNEDY, J. ARTHUR DE MONTELAMBERT, EDWARD TELLER**

12:30 PM The 1995 Conference Adjourns.

CONTENTS

CHAPTER III - THE NEXT FIFTY YEARS OF THE PEACEFUL APPLICATION OF NUCLEAR ENERGY

CHAPTER IV - NUCLEAR POWER GROWTH, NON-PROLIFERATION, AND POLITICS OF NUCLEAR ENERGY

CHAPTER V - INTERNATIONAL MANAGEMENT OF NUCLEAR POWER FUEL SYSTEMS

PRESENTATIONS ON THE INTERNATIONAL CONFERENCE ON ECONOMICS AND POLITICS OF ENERGY

José G. Martín

College of Mathematics, Science and Technology
University of Texas at Brownsville
Brownsville, Texas 78520

INTRODUCTION

The Global Foundation, Inc., a nonprofit organization for global issues and frontier problems in science, organized this International Conference around two closely related themes on which long-term energy strategies hinge: economics and politics.

At the conference, some of the world's most respected energy analysts shared their projections on global energy trends, technological changes, and fuel geopolitics. These projections provided the basis for the identification of foreseeable energy scenarios for the next half century, and allowed a discussion on the types of technical approaches and sociopolitical structural developments that should be nurtured now.

The papers reproduced later in these proceedings have been submitted by the participants themselves, and they represent a rigorous statement of the insights shared at the Conference. These informal notes, by contrast, represent a modest effort by this annotator to capture some of the highlights of the oral presentations and the ensuing discussions. The specific choice of the highlighted material is very much a subjective matter, and - *caveat vendor* - it is subject to the annotator's own misinterpretations. The material has definitely not been reviewed by the speakers. The only justification for this rather cavalier approach is that it may make it possible to preserve something of the excitement of the presentations, and possibly help leave a record of some of the material that came up which has not been included in the authors' own written submissions.

There were some notable absences at the Conference this time. Although it is not ever wise to use the word "indomitable", the temptation is hard to resist when referring to Dr. Edward Teller, of the Lawrence Livermore Laboratory. Resistance is well-nigh impossible when Dr. Teller's announced presentation is *"Misconceptions Connected With Energy Policies."* Unfortunately, Professor Teller had to miss this Conference - this was in fact his first absence from this series of "Global Conference." The Conference participants were also looking forward to listening to Dr. Glenn T. Seaborg, of the Lawrence Berkeley National Laboratory, who was to speak on "The Next 50 Years of the Peaceful Applications of Nuclear Energy", and Dr. John W. Landis, of Stone and Webster, who was going to present *"The Worldwide Perspectives for Nuclear Energy"*. Unfortunately, they could not attend either. We wish them all a speedy recovery - and remind them that they most bring a note from their doctor to the next Conference.

In these informal notes, any insights are clearly those of the participants, and any mistakes are this annotator's. It is hoped that these informal notes may serve to the conference participants as fresh reminders of some lively constructive discussions. Further, we may hope that it will help those, like Drs. Landis, Seaborg and Teller and Seaborg, who could not attend the Conference this time, to join vicariously in the fun. These notes take the place of one of those postcards with a view of Miami Beach and a note saying *"wish you were here."*

THE MEETINGS

SESSION I

GLOBAL ENERGY DEMAND PROJECTIONS IN THE COMING CENTURY, DRIVEN BY POPULATION, ECONOMICS AND ENERGY EFFICIENCY

Moderator and Organizer: Behram N. Kursunoglu, Global Foundation, P. O. Box 249055, Coral Gables, FL 33124-9055
Annotators: **Earlie Marie Hanson,** Los Alamos National Laboratory, New Mexico
J. Jeffrey Irons, Northrop Grumman Aerospace
Dr. Behram N. Kursunoglu, Global Foundation

"Equipartition of the Nuclear Fuel Cycle as an Alternative for Non-proliferation"

Dr. Kursunoglu opened the conference welcoming the participants, and bemoaning the absence of illustrious colleagues who could not be with us in Miami Beach. Dr. Edward Teller, whose presence had electrified the 17 earlier conferences in the series, had suffered a stroke. Ambassador Nelson Sievering had undergone surgery, and Dr. John Landis could not be present either. He thanked our absent colleagues for their unremitting support in the past, and wholeheartedly wish them the best in all of our names.

He reported on the last conference on "Global Energy in Transition", and announced that the Proceedings would be available in January of 1996.

The topic of Dr. Kursunoglu's introductory lecture was the concept of the "equipartition" of the nuclear fuel cycle as a means to forestall global proliferation while making the benefits of nuclear power available to all peoples.

Dr. Kursunoglu suggests that different components of the nuclear cycle could be made the responsibilities of different countries - countries which did not have to be necessarily friendly to each other.

He offered a scenario in which several 1000 Mwe nuclear power plant could be built at the border between Iran and Iraq, attached to the electric grid of both countries, which would share responsibility for the plants. Equipment, fuel, and services for such plants could be acquired from several nuclear energy exporting countries; for example, from Russia, U. S., and/or Great Britain. Five nations could share responsibility for the nuclear cycle, there could no misrepresentation, and it would be impossible to use these reactors to fabricate weapon-grade material.

The "glory" would be shared by all participating countries, and it would be possible to control the fissile material to avoid the nightmare of proliferation.

Dr. Kursunoglu recognized that the stigma now associated throughout much of the world with nuclear power and radiation constitute a major obstacle to the equipartition concept.

Two decades ago, nuclear energy was reasonably accepted, and its use to improve the human condition was not considered a sin. Now, we even involve in doublespeak to avoid using the word "nuclear". Patients are not helped by "nuclear magnetic (NM)" diagnostic techniques anymore - the technology is the same, but it has been renamed "magnetic resonance imaging (MRI)."

As an example of the distortion introduced by the stigma, Dr. Kursunoglu recounted that at the time of the Three Mile Island event, he had arranged an appearance of a distinguished group of scientists before the House Committee of Science and Technology to argue that this event should not be used as an excuse that prevented the nation to take advantage of the option afforded by nuclear power. The group included Hans Bohr and Eugene Wigner, and Edward Teller. As they waited for an opportunity to testify, about 10,000 people demonstrated against nuclear energy in the streets, led by Ms. Jane Fonda. Of course, the scientists were never received by President Carter.

Because of the stigma unfairly associated with nuclear power, a valuable tool to improve human welfare has *de facto* been given up in this country. U. S. actions have also affected the alternatives in much of the world, where the energy for development is been largely derived from fossil fuels.

However, the environmental impact of accelerated development based on fossil fuels is becoming more obviously unacceptable every passing day. Countries where rationality prevails, and notably those with successful economic policies, rely more and more on nuclear power. This is true for the countries in the Pacific Rim, and it is also true for a country like France, which has the competitive advantage provided by its nuclear plants. (Those plants provide 80% of the electric power generated in France.) The speaker argued that so long as rationality prevails somewhere, nuclear power will make a contribution to help meet human needs. Dr. Kursunoglu then proceeded to introduce the keynote dissertator.

Keynote Dissertation:

Dr. Richard Balzhiser, President, Electric Power Research Institute, 3412 Hillview Avenue, Palo Alto, California 94304-1395

"Global Energy and Electricity Futures"

Dr. Balszhiser opened his talk on electricity futures by stating that, in the U.S., those concerned with forecasting possible paths had to contend with the fact that gas was cheap here: its low cost limits the economic competitive of the alternatives in the U.S.

The U.S. will not dominate global energy demand for long, however: *"demography is destiny"*, and the large scale economic development and population expansion (8 to 10 billion people by 2050!) of the developing world will largely define global futures in the next century.

He emphasized three points: 1) the pressure on resources imposed by the expansion - giving China as an example; 2) the shared goal of sustainable development, with an emphasis on efficiency and taking advantage of advances in electric precision, control, versatility, information, and intelligence; and 3) the world appetite for coal, which will not decline.

He cited forecasts that predict that fossil fuel consumption would peak by the year 2025, while renewable and nuclear energy will predominate by the year 2060. Reviewing the availability of fossil fuels, he emphasized the uneven distribution of resources: coal is abundand in Asia, Russia, and North America; oil, in the Middle East, and gas, in the Middle East and Russia.

Our first priority, according to the World Energy Council, should be *"to relieve poverty in the developing world"*. Which technology is appropriate may differ from country to country. The "electricity fraction" of the energy mix differs greatly from developed to developing countries. Electrification is capital intensive, however, and efficiency should be the backbone of sustainability.

Dr. Balzhiser reviewed present trends, such as the evolution from a reliance on central station to an emphasis on modular distributed systems and the potential for natural gas plants to deliver "just in time" power. The speaker dwelled at some length on integrated coal gasification combined cycle plants (IGCC), with great potential in countries with vast coal reserves such as China: IGCC's are as controllable as gas fired plants but they are more expensive.

He reviewed the potential of photovoltaic (PV) systems to supply village power in countries like India, noting that PV cannot compete on a least cost basis for distributed systems. Wind power, potentially attractive in both Europe and India, is not as robust as pv, but is more economical: with technology from U.S. Windpower, at winds of 16 mph, it is possible to generate electricity at US$0.05 per kwh; the target, in 350 Kw turbines, is $0.03 per kwh, and this target is achievable in five years.

Nuclear power is the cheapest (predicted nuclear electricity costs average $0.022/kwh) but the viability of this option requires a *political* solution to safety concerns. The speaker also noted that, in addition to advanced light water reactors, we should also consider high-temperature gas cooled reactors

Throughout time, there is a change in the critical technology systems which demand electricity: in the 1880's, it was light motors; nowadays, the emphasis is on refrigeration and air conditioning, TV's and computers; in the future, it may be electric vehicles, electrotechnologies, heat pumps, etc.

There are three important opportunities for technological improvements: transportation (where changes may improve air quality); space conditioning, and industrial uses - where processes that require heat may be superseded by productivity improvements brought about by lasers and ultraviolet radiation.

The U. S., which led the birth of nuclear power, led the world in its decline, but the rebirth of nuclear power is imminent. There is an aggressive need for new reactors. Nuclear fission is the key to enable the world to meet its energy needs, and it is also the key to space exploration.

In the long term, we need nuclear fusion, which may based on magnetic or inertial confinement. There is a broad range of approaches for fusion - the neutron fusion reactor, colliding beams, etc., but the hurdles have not been overcome.

Nuclear energy can help meet human needs in many other ways: through radioisotope applications in space (SNAP sources), or in the manufacturing of highly sensitive gauges to inspect finished goods (manufacturing steel, aluminum, aircraft engines, pipelines). Nuclear medicine helps many: radioisotopes are involved in the diagnose or treatment of one third of all patients. Radioisotopes can be used to sterilize cosmetics. The use of fertilizers or insecticides can be reduced by using radio tracers, and nondestructive testing can be a valuable tool in arts and the humanities.

The investigation of transuranic nuclide is an exciting field - there are many potential applications for superheavy elements in postulated "islands of stability".

After surveying many fields where nuclear technologies can make a contribution, the speaker stated that any attempt to slow down the process through which we advance the state of the art in these technologies will slow down progress.

SESSION II

Energy Supply Projections of Primary Fuels, Renewables, Conservation, and Nuclear Power

Session Organizer: William M. Jacobi, Monroeville, PA
Moderator: Gerald Clark, Uranium Institute, London
Annotator: Lee Elder, Black and Veatch

Dissertators:

Dr. Steve Freedman, Gas Research Institute, 8600 West Bryn-Mawr Ave., Chicago, Illinois 60631

"The Role of Natural Gas in Electric Power Generation in the Twenty-First Century"

Dr. Freedman gave a cogent and comprehensive perspective on the likely role of natural gas in the next few decades. It would have been hard for a participant to ask, even rhetorically, whether there could be anything missing in such a perspective.

He started with a general discussion on increased global energy consumption and noted that, because of implicit institutional inertia of a type with which we are well familiar in the U.S.A., decisions being made today will affect the global environment for many decades.

He then proceeded to preempt any arguments from proponents of sustainable energy sources by conceding - in fact, by arguing - that eventually non-fossil energy sources will be needed. The catch, of course, are the near-term "practical concerns," which *de facto* limit all viable short-and-medium term power options to those based on fossil-fuels.

Once the options are thus narrowed, it follows that natural gas options, ranging from combined-cycle cogeneration plants to distributed modular fuel cells, are far superior to alternatives such as oil or coal, "straight" or gasified. Oil is too scarce, and coal is too dirty. The obvious spoiler is, of course, that we do not know how much gas is there available at competitive prices. Rather than ignoring this question, Dr. Freedman faces it head-on: he values it as the $3 trillion question, and answers it with persuasive arguments on the fallacy of Malthusian-like forecasts, on the power deregulation has in the creation of new gas reserves, and on the potential for technological improvements to overcome estimates which are by necessity influenced by our own cognitive limitations.

An important subtheme of the presentation was that the implied assumption of limited gas reserves world-wide is leading countries with expanding economies to choose energy options with a potentially large environmental impact, and that institutional inertia will make those options affect the global environment for many decades to come.

The paper represented a rather thorough and exciting review of gas-powered options and on the impact of the advances in controls and in materials on those options. Understandably, there was an emphasis on combined cycles and/or cogeneration, but there was also a discussion of gas turbines, distributed generation (including fuel cells for urban areas, gas turbines and reciprocating engines), with valuable information on costs, reliability, etc. In fact, the discussion was so comprehensive and well-reasoned that at times it seemed as though the speaker was doing the equivalent of beating of dead horse - by now it is clear that natural gas options have strong economic, environmental, and political advantages over the options, fossil or otherwise. The speaker gave the impression that we could do worse than spending sometime reviewing what we know about gas turbines.

In fact, the speaker covered so much ground, so thoroughly, that this annotator could find no fault of either commission or omission. Consider the statement "How much gas there is... is not known. Not known is not the same as known to be zero, even though zero is a possibility." What could possibly be wrong with this statement, other than for some carelessness in the use of probability concepts?

What happens if there is in fact little gas? Once again, the speaker followed the argument: *"Planning for the future must include the concept of a horizon beyond which additional information...will penetrate the market."* He also suggested that *"forecasts should be periodically updated to account for developments... which often evolve in unforeseen directions."*

(Note from JGM: At the risk of appearing to be a spoilsport, this annotator proposes that the time to start updating our forecasts is right now. Is it reasonable to expect that the gas reserves are enough to meet the exploding needs of the developing world? What should countries without gas reserves do? Even it we take for granted that there may be enough gas to meet conceivable demand in the U.S. in the next decade or two, is there a cloud to this silver-lining? Are there any advantages in diversifying our energy portfolio? What is the price that our country will have to pay if in fact there is modest room for further gas findings? How much time shall we be allotted to change the forecasts if we narrow our options now? Last but not least, what is the potential for gas scarcity elsewhere to spur technological and social breakthroughs? Will the U.S. Enterprise ship be gas-powered in a future Startrek TV Series?)

Mr. Raymond Sero, General Manager, International Operations, Westinghouse Energy Systems, P. O. Box 355, Pittsburgh, Pennsylvania 15230

"Potential Nuclear Energy Growth in Eastern Europe and Southeast Asia"

Mr. Sero made a comparative analysis of the prospects for nuclear power in countries in the former Soviet block and in Southeast Asia, emphasizing analogies and contrasts. The Eastern European markets reviewed included Bulgaria, the Czech Republic, Ukraine (there is active nuclear construction in these three countries), Poland, Hungary, Romania, and Slovakia, in addition to Russia. Indonesia and Thailand were highlighted as countries with little previous nuclear industry but with ambitious nuclear plans for the near future.

Some common characteristics for both groups of countries are that the economies are in transit, the countries are attracting new investments, and in both groups there are pressing environmental concerns and institutional challenges. Also, in both groups economic policies are now emphasizing privatization, restructuring costs so that they reflect costs, and dramatically changing the regulatory framework.

There are important differences, however. The most important difference, of course, is that Southeast Asian economies are booming and those of the Eastern European countries are not: gross national product is growing by rates of 7-1/2 to 8.3% per year in the former, while the latter may expect at most rates of 3-1/2 to 4% per year.

Southeast Asian countries are new to nuclear construction, while the Eastern European countries are not. On the other hand, in Eastern Europe, there is no leading enforcer of nuclear expansion plans and there is a "weak nuclear safety culture", while many Southeast Asia have a successful nuclear plan and a strong nuclear safety culture.
Eastern Europe depends more heavily on financing provided by the G7 countries interested in promoting political stability, while Southeast Asia has its own sources of capital, from commercial sources, private markets, and export credits.

Nuclear plants are going to make an increasing contribution toward helping meet expanding energy needs in both regions in the first decade of the XXI Century.

Randy Hudson, Oak Ridge National Laboratory, Oak Ridge National Laboratory, P. O. Box 2009, Building 9102-1, Oak Ridge, Tennessee 37831-8038

"World Energy Use- Trends of Demand"

Through graphs depicting past data from 1980, and trends projected onto the year 2010, the speaker gave a presentation on world energy use and projections by fuel type and region. The projections are derived from the Energy Information Administration World Energy Projection System.

For the period covered, far more energy is generated by oil burning than from any other source. Although more energy is consumed in North America than by any other region at this time, the projections indicate that consumption from the Far East and Oceania will surpass North American consumption before the year 2005.

Countries in the Far East and Oceania already release more carbon dioxide than any other, but their share will continue to increase in the foreseeable future.

SESSION III

ELECTRICITY GENERATION GROWTH IN FREE MARKET DEVELOPING COUNTRIES, WITH PARTICULAR ATTENTION TO CENTRAL VS. DECENTRAL SOURCES, RENEWABLES, AND COAL AND NUCLEAR

Moderator: M. John Robinson, Black and Veatch, 11401 Lamar, Overland Park, Kansas 66211
Organizer: Caulton L. Irwin, Department of Chemical Engineering, West Virginia University, P. O. Box 6064, Morgantown, West Virginia 26506
Annotators: **Dr. R. A. Krakowski,** Los Alamos National Laboratory, MS 607, Los Alamos, New Mexico 87545
Jean-Claude Guais, NUSYS; 9, rue Christophe Colomb, Paris 75008

Dissertators:

Gerald Clark, The Uranium Institute, London

"Nuclear Prospects in Southeast Asia and the Far East"

ABSTRACT

Driven by population and economic growth, world energy demand will double by 2020, but in the first fifteen years of this period world-wide nuclear capacity will expand by less than 20%. Unless plans change, the nuclear share of the world electricity market will decline in the next thirty years to about two thirds of its present level.

Electricity generation is the core business of the nuclear fuel industry. Although much of the planned nuclear expansion will occur in the Far East and South East Asia: a larger nuclear role in these countries is possible - and inevitable, if these countries are to abide by the resolutions of the Rio Convention on Climate Change.

In Europe and North America, largely because of the slow growth of electricity demand nuclear power is simply "marking time." In the great arc of territory stretching from Japan to Iran, the picture is different. There, conservation takes second place to increased supply, and there are 16 new reactors under construction in China, India, Japan, Pakistan and South Korea. Other countries such as Indonesia, Iran, Malaysia, Thailand, and Vietnam, which do not at present have nuclear power, have announced nuclear development plans.. The same is true for Turkey.

In most of the mentioned countries, electricity is largely generated from fossil fuels (coal in China, oil in Indonesia, Iran, Malaysia and Thailand.) In countries without oil reserves, imports adversely affect trade balances. Even in Iran and Indonesia, nuclear plants would improve the trade balance by releasing oil for exports.

As it became clear at the Berlin Conference on Climate Change last year, environmental concerns are also important. Also, in countries such as China and India, transportation and transmission costs give nuclear an advantage.

Regarding plans for nuclear construction, the Eastern Asian countries can be classified under three categories: those who long ago made the commitment to nuclear and already have a significant number of nuclear stations (Japan, South Korea and Taiwan); countries which have embarked on a nuclear program, making some progress (China and Iran); and countries which have only now reached the stage where it makes sense to contemplate nuclear power. In the last two categories, the nuclear plans do not represent a quick fix: they do nothing for the immediate needs - and not much for the medium-term ones.

Some interesting patterns have emerged. CANDU technology is playing a notable part in Korea and Pakistan. Korea is building more, China is assessing them; and so are Indonesia and Turkey. Indian technology was originally developed in conjunction with AECL.

Russia plays an active role in China and Iran. Russia and China have exploited the US-led boycott of Iran. Russia is building a centrifuge enrichment plant in China, and there is talk of another in Iran. The old technical alliance between Russia and India may be revived. China is building a reactor in Pakistan and hopes to sell further technology and services. South Korea is setting up deals using their basic design in North Korea, Vietnam and China.

Despite French successes in China, Western PWR technologies are not dominating the technology in those countries which still order new power reactors. The Korean basic design, although based on Combustion Engineering technology, may supersede its model.

The obstacles to the nuclear expansion are institutional/political and financial.

Politically, there is a tendency to assume that the spread of civilian nuclear power increases the risk of proliferation, even though this assumption does not match historical facts. The Non Proliferation Treaty has been very effective, and the non-signatory countries have found their nuclear progress more and more difficult. The Extension Conference has the long term result of making it easier for other countries to enjoy the benefits of nuclear power, and make them more inclined to apply safeguards rigorously.

Financially, limited availability of capital is in many cases the main limiting factor for nuclear power projects. Developing countries usually lack the internally generated funds necessary to finance nuclear plants costing several billions of dollars - they must raise the funds externally, and have the problem of a high level of indebtedness caused in part by previous power sector investments.

Possible solutions are utility financing, joint ventures, and project financing of the build-operate-transfer type. Joint ventures have been used successfully in China and are under consideration in Turkey.

Mr. Philip W. Covell, Mr. Richard D. Hansen, Enersol Associates, One Summer Street, Somerville, Massachusetts, and
José G. Martín, University of Massachusetts, Lowell, Massachusetts 01854

"The Role of Subsidies and Private Investments in Sustainable Rural Electrification"

The main points of this presentation were 1) that the widespread use of PV for rural electrification is desirable; 2) that full cost recovery is a key factor to ensure widespread use; 3) that temporary equipment and financial subsidies can hurt; 4) that developers and planners should adopt pricing strategies that reflect market conditions, and 5) that development assistance should focus on risk reduction through planning, technical assistance, training, and policy making.

The paper, reproduced later on in these proceedings, gave a global view of energy generation and consumption in transition, to justify the need for a diversified approach that includes renewable energy sources such as nuclear, conservation, and solar power utilization.

Reviewing the financial consequences of the ongoing shift on global energy patterns, it was argued that to ensure a diversified energy portfolio the means must be found to attract private capital to renewable energy sources, and in particular to solar rural electrification. To attract private investments, however, policies must be implemented to enable full cost recovery on the investments.

Well-meaning policies from governments and private voluntary organizations around the world often rely on equipment subsidies for the promotion of photovoltaic systems. A review of subsidized projects around the world helps to underscore what should be obvious: that these policies may actually discourage private investment, and thus work against the effective market penetration of renewable technologies. This paper debunks some of the myths associated with solar subsidies.

The goal of full cost recovery for solar investments implies a critical review of the role of subsidies, but this review must be undertaken in the context of the generalized use of subsidies throughout all the world for all kinds of energy alternatives without much regard to the environmental and social consequences of the alternatives. Everything from government-sponsored research in nuclear energy to World Bank loans for hydroelectric power, and from the Gulf War to artificially low kerosene prices for the rural populations in poor countries represent subsidies that distort the market. In this context, the goal of full cost recovery for solar rural electrification cannot imply a complete elimination of all subsidies, but a rational evaluation of those subsidies so as to assist effectively in ensuring market penetration for photovoltaic systems.

Rather than eliminating market interventions, the question should be how to structure those interventions to ensure long-term benefits and to mitigate risks for financial institutions, investors, and firms in the industry.

The presentation argued that, from the point of view of government and private voluntary organizations, development assistance should focus on risk reduction through planning, technical assistance, training, and policy making. Also, whereas temporary equipment and financial subsidies are not helpful for securing private resources over the long term, an emphasis on risk reduction in the form of initial infusions of project development support for efforts that feature full cost recovery can attract capital to make the photovoltaic technology broadly available.

A rational approach to market interventions will benefit rural people in developing countries - and help alleviate the formidable environmental and financial challenges facing all of us in the next few decades.

Kim Heinz, Southern Electric International

"Globalization of the Electricity Industry"

(notes missing)

SECTION IV

THE NEXT FIFTY YEARS OF THE PEACEFUL APPLICATIONS OF NUCLEAR ENERGY

Moderator: Dr. Bertram Wolfe, Vice-President, GE Nuclear Division, 15453 Via Vaquero, Monte Sereno, California 02138
Organizer: Dr. Edward Arthur, Los Alamos Scientific Laboratory, P. O. Box 1663 - MS F628, Los Alamos, New Mexico 87545
Annotators: Dr. Marcelo Alonso, Florida Institute of Technology, 509 Third Avenue, Melbourne Beach, Florida 32951
Dr. John W. Zondlo, Department of Chemical Engineering, West Virginia University, P. O. Box 6102, Morgantown, West Virginia 26505
Dr. John Ireland, Deputy Director, TSA Division, Los Alamos Scientific Laboratory, Mail Stop F606, Los Alamos, New Mexico 87545

Dissertators:

Dr. Linden Blue, General Atomics Corporation, P. O. Box 85608, San Diego, California 85608

"Nuclear Power in its Second Half Century"

To illustrate how really small and thus vulnerable our atmosphere is to contamination by gas emissions, Dr. Blue asked the audience to imagine a basketball in Saran wrap - in comparison with the ball, the wrap is as thin as the atmosphere is to the Earth. To illustrate how vulnerable our fresh water supply is to liquid discharges, he asked the audience to think of the world as an orange - and to fresh water as a drop.

Nuclear makes an important contribution to our needs: it makes possible the saving of the fossil fuel equivalent of all the cars in the world. It saves U.S. consumers the equivalent of $ 1 billion/week of imported oil, representing over one half of our international balance of payments.

For the next half century, we may want to repostulate the goals of nuclear power development. If safety is an issue, we may want to go to lower power densities. If one wants to make it impossible to have a core meltdown, one should use materials such as graphite or ceramics. Better materials and high burn up may be the key to less dangerous wastes, and if corrosion is a concern, one could use a non-corrosive fluid like helium - or get rid of steam generation entirely.

We have developed such reactors, with low power densities, high temperatures, and ceramic fuels. (Peach Bottom, etc.) The reactor is modular, and the conversion equipment is a gas turbine.

Dr. Shelby Brewer, Chairman, ABB Combustion Engineering, Inc., 1000 Prospect Hill Road, Post Office Box 500, Windsor, CT 06095-0500

"Nuclear Power in Its Second Half Century"

The fact that our nuclear plants are powered by light water reactors is a historical accident which resulted from the work on nuclear submarines. The U. S. government was partner to the nuclear industry in its first half century, but it has not been a reliable partner, and its embrace has been fateful.

Macroeconomics and politics will drive nuclear development. In the interreign, the commercial structure will be reshaped. Who will own the technology? Will there be a nuclear monopoly, like in France? Shall we retain the status quo , where the utilities absorb the risks, or will the vendors accept more risk? Shall we have turnkey projects, where the risks and profits are made by the suppliers, and shall we have nuclear independent power producers?

There has been attempts at turnkey projects. For the Taiwan Nuclear Project, there was an RFP 1-1/2 year ago, based on the "total turnkey" concept. No proposal came under the budget. Last April/June, the project was opened for new bids, and there were responses from GE,ABB, Framaton, and Westinghouse.

For nuclear power to fulfill its role in meeting national energy needs, it requires a less obtrusive, more dependable partner, and one which can fulfill statutory requirements. The next era poses major challenges, and institutional, financial, and political hurdles.

Dr. Poong-Eil Juhn, Director of Nuclear Power, Department of Nuclear Energy and Safety, International Atomic Energy Agency, Wagramerstrasse 5, A-1400 Vienna, Austria.

"The Worldwide Perspectives for Nuclear Energy"

Dr. Juhn reviewed the contribution that nuclear power makes to meet global energy needs, and outlined an optimistic perspective on the near-term future. He quantified his outline in terms of fuel savings and in terms of avoided greenhouse emissions.

William D. Magwood, Associate Director, Office of Planning and Analysis, U. S. Department of Energy, NE-1/5A-143, 100 Independence Avenue, SW, Washington, D. C. 20585

"Department of Energy Perspectives on the Future of Nuclear Power"

Dr. Magwood addressed the issue of the future of the nuclear industry; he encouraged the nuclear power utility to support university nuclear programs, and encouraged nuclear educational programs to make students have a historical perspective of nuclear energy. He emphasized the role of nuclear power in limiting atmospheric effluents, atmospheric chemical activity, and acid rain. He also highlighted plant extension: *"everybody talks about extending plant life by twenty years."*

Dr. Don W. Miller, Professor and Chair, Ohio State University, Nuclear Engineering Program, 1079 Robinson Lab , 206 West 18th Avenue, Columbus, Ohio 43210-1107

"The International Nuclear Societies Council Vision of the Second Fifty Years of Nuclear Science and Technology"

Dr. Miller shared the International Nuclear Society Council Vision of the next fifty years of nuclear science and technology. He picked up the team of the quality of life in developing countries, and stated that quality is dependent on three closely intertwined elements: adequate energy, food supply and medical care.

Short of a disastrous global epidemic, the world population will double by the middle of the XXI Century, and instantaneous communications via television and internet have raised the expectations of that population for a better life.

A better life depends on the efficient use of resources and energy. The citizens of the wealthy countries today are healthier, better educated and live longer than any other people at any other time in the history of civilization. Even if environmental concerns are not a high priority in most of the world, those concerns will eventually dictate the way of life world wide.

We can look at energy demand in the next 50 years from two points of view. We can base our forecasts on the need to improve the quality of life in the world at life. There are indicators which suggest that quality of life could be improved by increasing consumption in developing countries to about half the current use in the U.S. and Canada, or 100 Gigajoules per capita. Alternatively, we can take a historical view. Energy demand has increased at 2.3 % per year for the past 130 years: maintaining that growth over the next fifty years would lead to the same global consumption.

There is no reason why such energy use is not possible - if there is a will at the political level. On the issue of food supply, a key requirement is distribution. On the issue of energy, this requirement can be met if current high energy consumers share their technology, do not limit energy use in the developing countries, and curtail their own rate of increase of energy use.

Dr. Miller shares the results from models that suggest that gas and nuclear will be the dominant energy sources in the next 50 years.

If today's mix of energy sources was maintained, there would be a three-fold increase use of nuclear over the next 50 years. However, if the Rio Convention agreement is to be met, the demand for nuclear can only be met by a twenty-fold increase.

To compete, there must be changes. At present, capital costs are higher than anticipated because of expensive delays and changing regulatory requirements.

The speaker outlined steps that could facilitate nuclear power expansion, and then proceeded to elaborate on the use of nuclear technology for medical therapeutic and diagnostic uses, the preservation of food, etc.

Many more poor people could be adequately fed if we were not subject to paranoia by self-interest groups whose real agenda is more social change than public protection.

According to Dr. Miller, we have the knowledge and expertise - and the moral obligation to ensure that nuclear technology is available for use by the underprivileged majority as a means to substantially reduce their real risks, which are lack of food and lack of energy.

We must provide the leadership to change the regulation of nuclear technology such that it properly reflects current knowledge and understanding of health effects. We must design standardized and easily manufactured nuclear power plants. We must continue our efforts to educate the people on the benefits of preservation of food by radiation and the generation of energy by nuclear fission.

And we cannot stand by and permit a minority which is subject to minimal in risk in their own lives to inhibit the use by the underprivileged majority of a technology that will substantially reduce their real risks and improve their quality of life.

SECTION V

NUCLEAR POWER GROWTH, NONPROLIFERATION, AND POLITICS OF NUCLEAR ENERGY

Moderator: Ambassador Richard Kennedy, 2510 Virginia Avenue, Northwest, Washington, D. C. 20037
Organizer: Dr. Edward Arthur, Los Alamos Scientific Laboratory, P. O. Box 1663 - MS F628, Los Alamos, New Mexico 87545
Annotators: Dr. R. A. Krakowski, Los Alamos National Laboratory, MS 607, Los Alamos, New Mexico 87545
Jean-Claude Guais, NUSYS, 9, rue Christophe Colomb, Paris 75008, Paris

Dissertators:

Mr. Harold Bengelsdorf, 6510 East Halbert Road, Bethesda, Maryland, 20817

"The Special "Blue Ribbon" Panel Report Issued by the American Nuclear Society on the Protection and Management of Plutonium"

The Special "Blue Ribbon" Panel, with Dr. Glen T. Seaborg as Honorary Chair, issued a report in 1994 for the National Academy of Science on how to dispose of excess Plutonium, and identified several options, ranging from immobilization to conversion to mixed oxides (MOX) which could be burned in LWR's or CANDU reactors.

Dr. Phillip Sewell, Vice-President for Corporate Development, U. S. Enrichment Corporation, 6903 Rockledge Drive, Bethesda, MD 20817

"U. S.-Russian HEU Agreement"

Dr. Sewell spoke on the Agreement to convert weapons material for peaceful purposes, under a 20-yr, $12 billion agreement. Under the Agreement, $60 M have already been advanced to lease warheads from Ukraine, convert the material into reactor fuel, and send it back to Ukraine. Six million tonnes of highly enriched uranium, from 250 warheads, represent enough fuel for 16 reactors, providing enough energy for ten million households. Dr. Sewell reviewed considerations on disarmament, nonproliferation, and an antidumping suspension agreement, to forecast a bright future for the Agreement.

Mr. Joseph D. Lehman, Director of Business Development, Martin Marietta Astronautics Group, D. C. 1130, P. O. Box 179, Denver, Colorado 80201

"Intelligence Requirements and Non-Proliferation"

The speaker presented an informative analysis of intelligence goals and means, in the context of the threat of nuclear proliferation.

First, he reviewed the reasons why there is a threat, that is, why different countries may be interested in acquiring nuclear weapons. The reasons included the presumption of prestige associated with "belonging to the nuclear club", the belief that nuclear capability provides a means of deterrence (the speaker mentioned Israel as an example of this driving interest); the assumption that the capability may serve as a leverage in international negotiations, and finally the search for the ability to inflict revenge on a potential aggressor.

He outlined the challenge posed by terrorism in the MidEast, South Asia, Korea, South America, Japan, and Germany.

Mr. Lehman then proceeded to analyze what the possible objectives of an anti-proliferation program would be: "obviation" (that is, the elimination of incentives); dissuasion, interdiction (that is, physical intervention on transfer and training); preemption (such as the attack on the Iranian reactor in the 80's); and retaliation.

Next, he listed sources and methods. These included

"IMINT", or image intelligence, (the U-2 spy planes; the images of the Cuban missile crises);
"ELINT", or electronic intelligence;
"SIGINT," or signals intelligence (radar);
"COM/INT", or communication intelligence;
"HUMINT", or human intelligence and espionage.

Under espionage, he considered political and industrial types, commercial sources, and diplomatic exchange. An anti-proliferation program requires diplomacy, enforceable treaties, regional alliances, and programs enhancing energy security.

The speaker reviewed those technical areas where progress is made or needed (affordable multispectral collection, enhanced signal and imaging processes, improved data fusion, etc.), and noted the changing intelligence environment created by the Internet and the World Wide Web.

J. Arthur de Montalembert, Cogema, France
(Paper co-authored by M. McMurphy, of Cogema, Inc., U.S.A.)

"The Civilian Recycling Industry: A Key Contribution to Disarmament"

The civilian plutonium recycling industry has accumulated an operational experience with MOX fuel which will prove helpful for a sensible solution to the weapons plutonium disposition issue. Such a solution has been identified as a favored strategy in the conclusions of several recent committee reports in Europe and the U.S.A. analyzing the weapons plutonium disposition program

In a recent report, the National Academy of Science panel "judges that with a prompt decision to proceed in this direction, and given a high national priority assigned to the task, fabrication of weapons plutonium fuel could begin in the United States as soon as the year 2001.

The ANS Blue-Ribbon Panel concluded in a report released a few weeks ago that the use of plutonium as fuel in current commercial reactors is already taking place routinely in several countries and can, therefore, be implemented with little delay.

And a few days ago, at the Winter American Nuclear Society Conference, Nobel laureate Glen Seaborg, in a major presentation, strongly supported the irradiation of excess weapons plutonium in MOX fuel.

In addition, the Russian authorities have clearly stated their own determination to recycle their excess weapons plutonium and have already started preparations to that effect with the technical support of COGEMA and the French Atomic Energy Commission.

In Europe, reprocessing and plutonium recycling have attained industrial maturity and provide the foundation for a coherent European strategy. The utilities involved in MOX reactor operation and the fuel cycle industry experience no technical difficulties and the European plutonium recycling programs are growing steadily. At present, there are 20 reactors loaded with MOX fuel, and this number is expected to grow up to 50 around the year 2000. At that time, 25 tons of plutonium will then be used in MOX fuel each year to produce the electricity equivalence of 25 million tons of oil. The European strategy is coherent.

Such an industrial experience gives a unique reference for weapons plutonium disposition through MOX use in reactors, which appears as the option which is "technically safe, economically competitive, and non-proliferation efficient."

The rationale of weapons plutonium recycling as MOX fuel in LWRs is based on technical advantages, industrial availability, natural resource conservation, economy, and nonproliferation issues.

On the technical and industrial aspects, the speaker emphasized that the thermal power from MOX fabricated from weapons plutonium is about 7 to 20 times lower than from commercial plutonium MOX and that radiation protection issues are facilitated because of the reduction in alpha, gamma, and neutron activity. From the point of view of resource conservation, recycling 50 tons of weapons plutonium provides almost as much electricity as is generated per year in France, and saves the equivalent of 50 million tons of oil. From the point of view of economics, the recycling of weapons plutonium for electricity generation is the only alternative which provides significant economic returns. Furthermore, the cost of the weapons plutonium fuel is comparable, or lower than, the usual uranium oxide fuel.

The environmental advantage arises because utilization as MOX fuel halves the inventory of existing plutonium in a single recycling. Finally, from the point of view of non-proliferation concerns, the ratio of fissile plutonium to total plutonium in the MOX fuel is much lower than the ratio in uranium oxide.

The presentation included a review of the European commercial recycling experience, the reactor cores loaded with MOX fuel, the EDF Program for loading 900 MWe PWR's with MOX fuel, and the European strategy towards Plutonium balance. The speaker summarized the objectives of COGEMA MOX fuel assembly fabrication and postulated alternate "smooth" and "accelerated" scenarios for recycling.

In conclusion, M. de Montalembert stated that plutonium recycling in MOX fuel is an industrial reality in Europe and that it represents a major contribution to non-proliferation objectives. He volunteered that the European industry stands ready to contribute to similar weapons plutonium disposition programs in the U.S., offering the transfer of technology and know-how, technical assistance and engineering services, and assistance in possible "quick start" MOX fabrication programs.

SESSION VI

INTERNATIONAL MANAGEMENT OF NUCLEAR POWER SYSTEMS

Moderator: Mr. Harold Bengelsdorf, 6510 East Halbert Road, Bethesda, MD 20817
Organizer: Mr. Anthony J. Favale, Deputy Director - Energy Systems, Northrop Grumman Corporation, Mail Stop B29-25, 1111 Stewart Avenue, Bethpage, NY 11714-3588
Annotator: Dr. José G. Martín, University of Massachusetts Lowell, MA 01854

Dissertator

Dr. Gordon Michaels, Oak Ridge National Laboratory, P. O. Box 2009, Building 9102-1, Oak Ridge, TN 37831-8038

"Thermal Issues in U. S. Government Repository and the Potential Role for Waste Transmutation"

Dr. Michaels reviewed hydrological and thermal issues in the proposed U.S. Government Repository in Yucca Mountain. According to Dr. Michaels, those issues threaten to derail the U.S. plan for the Geological Repository: *"Where we are, is worse than the worse-case scenario"*.

The main issues seem to be the heat buildup in a non-vented depository, where long-range integrity depends on underground water not reaching the waste. This buildup induces a massive perturbation of the geologic site, creating flow cells which puts water in contact with the containers for the spent fuel at sites which were originally dry. In the hot and humid environment, corrosion will ensue, there will be containment breach, and the radioactive material will dissolve in the groundwater. Estimates of the rates for aqueous corrosion for high nickel allows illustrated the effect of temperature on container integrity.

The speaker explained the composite porosity model, differentiating between "old" and "young water" deposits and flows, and how the model predict the formation of crack systems which make spent fuel corrode much faster than originally predicted at certain points. The cracks act as heat pipes. The model predicts the eventual total collapse of the volume of rock above the spent fuel after a certain time which depends on the amount of thermal heat generated in the repository - and thus on the quantity and nature of the stored material. In the scoping calculations, this collapse will occur after 1300 or 5000 years.

The fraction of the total generated heat which originates in the actinides increases with time. At first, the contributions from fission products such as cesium and strontium dominate, but after 1000 years, Plutonium accounts for 99% of the heat.

There are many advantages to process the waste so as to eliminate the actinides, using them in MOX fuel or in some other fashion converting them to fission products. Reference was made repeatedly to M. Wilson's SAND 93-2675 document on the total system performance of the Yucca Plant.

(Note from JGM: The presentation reviewed many alternatives and examples of past experiences, ranging from the Swedish experience in granite deposit to salt repositories. It also included provocative statements and suggestions to the nuclear industry. In this annotator's judgment, they were provocative enough that it is the better part of valor not to attempt to summarize or interpret them here.)

Dr. Richard Wagner, Jr., Vice-President and Chief Scientist, KAMAN Sciences Corp., Washington Operations Division, 2560 Huntington Ave., Alexandria, VA 22903

"Global Plutonium"

Dr. Wagner praised the ANS Blue Ribbon Report on Plutonium, which has an impact on this talk on a global nuclear material vision.

He reviewed the strategies for minimizing global inventories, using the fissile material in critical reactors or in subcritical systems, using accelerators. Why is a large inventory of Pu or HEU dangerous? The speaker asked the rhetorical question, "As inventories are brought down, does the proliferation risk diminishes?

The current U.S. policies on reprocessing are counterproductive. Still, a blanket endorsement of a Pu-based economy is unwise. Is there a middle ground?

He explored the proliferation challenges in terms of different scenarios: the subnational threat, the canonical nth nation proliferation, the emergency of a near peer competitor, and widespread, large-scale, n-sided, rapid competitive nuclear rearmament.

The effectiveness of an anti-proliferation measure depends on the scenarios: in the short-range, lowering the weapon material inventory is irrelevant to the problem of the potential leakage of a few kg of Pu to a potential proliferator.

Comment by **Dr. Bertrand Wolfe**: *"One motivation for proliferation, and a reason why somebody wants weapons and may use weapons, is the lack of energy sources. What is needed is access to energy - we should utilize nuclear power, to minimize the danger."*

Dr. C. Pierre Zaleski, ESIC 4, (Université de Paris, Dauphine) F-78350 Jouy-en Josas, France.

"Management of Russian Military or Weapons-Grade Plutonium"

The policies being advanced in the U.S. to deal with excess weapon material from Russia are too narrow, and have a major weakness, made obvious in November of last year in Venice: they are not supported by Russia or France.

For a policy to be acceptable to Russia, the economic value associated with the plutonium must be as good as possible. Mixing Pu with fission products is not consistent with Russian policy, and is questionable from non-proliferation point of view.

A better possibility is through medium-cooled fast reactor.

Such a specially designed fast reactor would have a low specific power density, very long fuel residence time (10-20 years), high breeding ratios (in range of 0.9 to 1); and it would have no blanket. The idea would be to "store the Pu in the reactor - and keep it."

Fuel rod diameter would be large (more than twice that of present breeders.) It would have a simple vessel closure and fuel handling system. In such a reactor, Pu would be degraded.

A redesigned reactor with a re optimized core would be cheaper than the BN-800.

The speaker outlined the potential deployment and development of the fast-reactors, emphasizing the financial aspects. Some risks are technical (this would be a new project); there is also the risk of potential political and economical instability in Russia. However, the very existence of such a program would be helpful in securing stability.

SESSION VII

RELEVANCE OF INTERNATIONAL CONSENSUS POLICIES ON ALTERNATIVE NATIONAL ENERGY STRTEGIES AS RELATED TO FREE MARKET ECONOMIC, GLOBAL FUELED TRANSPORTAITON SYSTEMS FOR OIL AND GAS, ENVIRONMENTAL IMPACT AND GEOPOLITICAL TRENDS

Moderator: **M. Jean Couture,** 8 Rue du Jour, 75001 Paris, France
Organizer: **Dr. Carl W. Myers,** Los Alamos National Laboratory, Mail Stop D446, Los Alamos, New Mexico 87545
Annotators: **Mr. Lawrence F. Drbal**, Black and Veatch, P. O. Box 8405, Kansas City, Missouri 66211
Thomas Cunnington, Cyton Corporation, 2065 West Maple Road, Suite C301, Walled Lake, MI 48390

Dissertator:

Dr. Dan Hartley, Vice-President, Sandia National Laboratories, Organization 600, MS 0724, Alburquerque, NM 87185-5800

"Renewables: A Key Component of Our Global Energy Future"

Los Alamos National Laboratory is making important contributions in nuclear and fossil power generation. Because those options are being explored in detail by others in this Conference, this talk concentrated on renewables.

Dr. Hartley gave a short overview of the global energy picture, emphasizing the emergence of developing countries and in particular China. He reviewed some of the drivers on the demand-side, and some of the factors affecting the global picture: economics, financial limitations, energy dependence, international arrangements, environmental, international security, infrastructure, physical knowledge, sustainability, and social.

He compared fossil, sustainable, and renewable alternatives.

The economics for fossil sources was favorable and the infrastructure was favorable in developed countries while mixed in developing ones. The speaker rated the option as "uncertain" in terms international security, and unfavorable in terms of sustainability.

For nuclear, the economics was potentially favorable although there were regulatory implications in the U. S. The option enhanced national energy independence, and it was favorable in terms of international agreements. The infrastructure was fair, in spite of the issue of public fears. As far as sustainability, the option received a "mixed" rating.

Renewable energy sources, with the exception of hydropower, rated "unfavorable" in terms of economics and relatively favorable in terms of energy independence. International agreements favored the choice because of concern for greenhouse gases, but "footprint" was a problem. The infrastructure rating was "unfavorable" while (of course) sustainability was favorable.

Renewables are likely to play a relevant role. PV already enjoys favorable market niches. It is predicted that almost 50% of all energy will be based in renewable sources in 50 years; the fossil energy contribution will continue to expand, but will have a downturn around 2020.

Dr. Curt Mileikowsky, Doctor of Technology, Sweden (Avenue de Rochettaz 14 A, 1009 Pully, Switzerland)

"On Global Warming, Future Living Standard and Environment"

Referring to the increase in the CO_2 in the atmosphere, Dr. Mileikowsky suggests that the interpretation of the impact of this increase, and conflicting views on what actions, if any, should be taken to mitigate that impact, may be interpreted in terms describing a potential *environmental war*.

The amount of CO_2 in the atmosphere has increased from 2180 billion tonnes at the beginning of the industrial era to 2,800 billion tonnes now. According to the United Nations Intergovernment Panel on Climate Change (IPCC), anthropogenic emissions are the main cause of ongoing global warming. (Dr. Mileikowsky warned that this conclusion is met with skepticism by some experts.) About 23 billion tonnes per year of CO_2 are emitted per year, and of this quantity, 52% is taken by the oceans, land vegetation and the soil, and the rest (11 billion tonnes) stay in the atmosphere.

According to most predictive models, if global warming occurs, the effects are likely to be benign in some regions of the world - for example, Northern Russia and Siberia, and Northern Canada, are likely to benefit from a warmer climate. On the other hand, Northwestern Europe is likely to become much cooler. The temperature there may drop from 8 to 10 C, approximating the climate of Greenland, and giving London a climate like Spitzbergen. (Scientific American, November issue.)

The steps to be taken to mitigate CO_2 emission will depend on whether the sources of the emission are large power plants or distributed sources. For the former there is the option to capture the CO_2 and compress it, transport it and dispose of it in the deep ocean, or in depleted oil or gas wells, or in aquifers. (At high pressure, say, 3000 m under the surface, CO_2 is 3.75 times heavier than water, and sinks further.)

The speaker summarized the capacity for storage as follows:

Capacity for storage

Disposal	Storage capacity (Gigatonnes of Carbon)
deep ocean	20 million
exhausted oil wells, yr.	125
aquifers (below 800 m)	87
near surface terrestrial	very high
global forest regt.	50 - 100

The speaker presented the costs of these procedures for a 500 Mwe natural gas combined cycle with and without 85% CO_2 capture (The costs includes compression, transportation, dispersal.)

Cost increase mills/kwhe

efficiency w/o w/ capture	power cost w/o CO_2	CO_2 capture	compression	transportation deep aquifer	well	disposal	total
52	40.6	35	18+2 per 100 km	+0.1	+1.5 100 km	+1.6 +0.1 per	56 to 62

The speaker referred to new processes such as absorption by novalthonolamine, and quoted estimates by Koide at al., 1992, Riemer and Ornerod, IEA (1994), dominated by capture of C02.

The challenge posed by numerous small units such as cars, home heating systems, motorcars, etc., needs different solutions. In some cases, substitution by electric units may be a solution. For cars, the lack of good battery technology is an obstacles: hydrogen or methanol may be used as alternative fuels., but both of these are more expensive than gasoline.

The CO_2 effects may be mitigated by sequestering through "afforestation" (sic.) Dr. Mileikowsky discussed potential and costs of "afforestation". The alternative of short-rotation cropping for biomass production to replace fossil fuel, using fast-growing trees harvested every 3 to 10 years is more expensive than long-term sequestration but is more "sustainable." Forestation seems to be a relatively cheap sequestering method, but a key issue is that it requires very large areas, which are scarce in the industrialized or newly industrialized countries, but which may be found in the tropics. Thus a potential conflict arises.

The speaker discussed the potential of nuclear power and renewable energy sources to help alleviate the carbon emissions. For nuclear power to be a viable alternative, it is necessary to address the issue of technological stigmatization. This may involve technological solutions, and he suggests that these solutions would probably raise nuclear power costs.

Solar energy utilization may be helpful, but from cost and land use considerations it appears that the best potential is to cover roofs and southern facades with photovoltaic panels. Although this could not the whole solution in industrial countries, approaches such as photovoltaic assisted lighting for load management would reduce the need for other power sources.

The speaker discussed solar power generation in space in some detail, sharing very interesting information on material sourcing for space solar plants: apparently, 95% of the materials needed could be mined in the Moon.

In conclusion, the speaker noted that it is possible to avert possible catastrophes associated with carbon emissions, but that a rise in energy costs cannot be avoided. Unfortunately, higher energy costs may have a disproportionately important impact on developing countries economies. Further, since the effects from global warming differ for different regions of the earth, conflicts arise because it is not clear that those in favored regions would be willing to pay to help others in disfavored regions.

The data which is being collected will make it possible to verify or contradict the findings of the IPCC within ten to fifteen years. If the findings are confirmed, and no mitigation action is taken because of the costs and the potential conflicts, we may be facing climatic change and a major world crisis.

SESSION VIII

ROUND TABLE ON PROBABLE WORLDWIDE ENERGY SYSTEMS

Moderator: Dr. Henry King Stanford, President Emeritus of the Universities of Miami and Georgia, P. O. Box 1065, Americus, Georgia 31709

Dr. Steve Freedman, Gas Research Institute, 8600 West Bryn-Mawr Avenue, Chicago, Illinois 60631
- Gas will play a key "bridge" role in the next few decades, until sustainable energy sources can overcome financial and political hurdles.

Dr. Gerald Clark, Secretary-General, The Uranium Institute, Twelve Floor, Bowater House, 68 Knightsbridge, London SWIX FLT, U. K.
- Nuclear power will have to take an expanded role if we are going to "make a dent" or overcome the obstacles to our meeting the terms of the Rio Treaty on greenhouse emissions.

Dr. Bertrand Wolfe, Vice President, (retired) GE Nuclear Division, 15453 Via Vaquero, Monte Sereno, California 02138
- A prevailing attitude among some people trying to influence public opinion is to criticize everything, and reasoning seems to make no difference.

Ambassador Richard Kennedy, 2510 Virginia Avenue, Northwest, Washington, D.C. 20037
- Our planning horizon is short, and upheavals may result from energy shortages and environmental conflict.

Dr. C. Pierre Zaleski, ESIC 4, Impassee du Docteur Kurzenne, F-78350 Jouy-en-Josas, France
(Paris, retired)
- Forecasts are difficult. Nuclear power expansion is likely to be driven by the Fat East: France is already saturated. Most electric power derives in France is generated in nuclear plants, and this has not resulted from low interest rates. The key is a regulatory climate which, like that in France, should not be lax but should be stable.

Dr. Poong Eil Juhn, IAEA, Wagramerstrasse 5, A-1400 Vienna, Austria
The future of nuclear power is bright, particularly in Asia. Two key elements to support nuclear expansion are improvements in safety and the Non-Proliferation Treaty. A similar convention is needed to harmonize policies on nuclear waste.

Other issues discussed at the round table were the shortage of power and water, the near-term potential for desalination using nuclear reactors in the Middle East; long-term issues in nuclear safety; the study financed by the Sloane Foundation on the future of the nuclear industry.

There was a lively discussion in all of these issues, but as the Conference drew to a close, there seemed to be a consensus from the participants at the round table that a major global shift is under way, and that nuclear energy is going to be powering the awakening economies in the developing world. So will renewable energy, with gas playing a bridge role.

Politics and economics are going to affect this development, and in the future a changed global hierarchy may make it easier for different countries to improve the standards of living of their people. The successful countries are going to be those where policies are decided on the basis of fact and not irrational prejudices. A type of back-handed compliment was given to charismatic actress/activist Ms. Jane Fonda, as participants referred to the "Jane Fonda Syndrome" as a factor which needed no explanation. A typical comment was that "what the U.S. does with its energy policies is a domestic issue, but it would be irresponsible to give in to the Jane Fonda syndrome and to let nuclear power die here"

A Final Note from the Annotator

José G. Martín

Risking redundancy, the annotator wants to emphasize that these notes are extremely informal. It should not be assumed that they accurately represent the true contributions of the presenters, who had not way to review them and provide feedback. The papers that follow in these Proceedings are the right source to search for the valuable insight afforded by those contributions. and these notes are simply the observation of one participant, colored and weighted by the intermittence of his own interests and attention span. This participant appreciates Dr. Behram N. Kursunoglu's willingness to include these notes in the Proceedings, and he is grateful for the opportunity to share a memorable meeting in such distinguished company.

CHAPTER I

GLOBAL ENERGY DEMAND PROJECTIONS IN THE COMING CENTURY, DRIVEN BY POPULATION, ECONOMICS, AND ENERGY EFFICIENCY

GLOBAL ENERGY AND ELECTRICITY FUTURES

Richard E. Balzhiser
President and Chief Executive Officer

Electric Power Research Institute
Palo Alto, CA 94303

INTRODUCTION

It is impossible to look out into the next century without being struck by the looming presence of what we now call the "developing world." It gives meaning to what Ben Wattenberg said, "that demographics is destiny." Over the next half century, these nations will undertake an unprecedented, large-scale expansion in the use of commercial energy in their efforts to meet the basic needs of a rapidly growing population.

In my remarks, I want to emphasize three points. First, the fact that the drive of these developing nations for improvements in health, education, and welfare -- as well as some measure of economic parity with the industrialized nations -- will lead to inevitable pressure on resources, and the global environment, including climate.

Second, new means for global sustainable development will be required, with efficiency forming the backbone of all future strategies of sustainability. In this context, electricity has a crucial role to play. Its ability to bring precision, control and versatility to nearly every task, coupled with its ability to capture and convey information and intelligence, are central ingredients in the quest for 21st century productivity and efficiency.

And third, the world's prodigious appetite for energy will rest upon what I call the big three: the coal, renewable and uranium resources of the world. I'll have more to say on the resource outlook in just a moment, but first let's back up and look at the social drivers.

POPULATION

There are two basic drivers of energy use: population and economic development. In Figure 1 you can see that population growth is exploding -- from one billion in 1800... to two billion in 1900 ... to five and half billion today... to perhaps eight to ten billion by the middle of the next century. Much of this growth will occur in the developing nations where economic development is a top priority, and this development will require prodigious amounts of energy.

Economics and Politics of Energy
Edited by Kursunoglu *et al.*, Plenum Press, New York, 1996

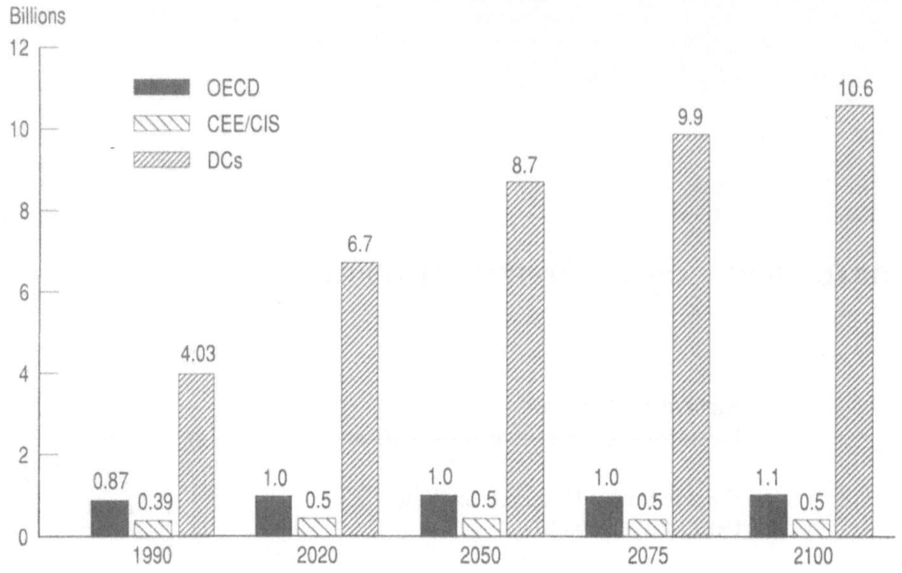

Figure 1. World Population: Actual and Estimates to 2100 by economic group. *Source:* World Energy Council.

WORLD ENERGY USE

World energy use will continue to soar, approximately tripling, by this projection, during the first half of the 21st century. This is the work of the Shell Planning Group (Figure 2), and assumes a net global increase in energy consumption of 2% per annum . This is fairly consistent with most of the mainstream forecasts. Where I find this particular forecast most fascinating -- especially given its source -- is that it shows fossil fuel use peaking around 2025, and the combination of renewables and nuclear accounting for over 50% of the world energy supply by 2060.

I happen to believe that coal will probably not decline as the Shell Group anticipates, because of its sheer abundance and concentration in Asia, and because of the role that technology can play in converting coal into a much cleaner, more acceptable and versatile fuel.

GLOBAL ENERGY RESOURCES

Virtually everybody agrees that much of the energy growth in the next quarter century is going to be fueled by coal, oil and gas, which as you can see are very unevenly distributed around the world. Coal is the overwhelming resource in Asia, Russia, and North America. The only significant supply of oil is in the Middle East. And gas is particularly abundant in the Middle East and the Russian territories. Significantly, exploration and drilling technology has made major contributions to increasing the reserve-to-production ratios for both oil and gas in recent years.

Nevertheless, you can see the disparities (Figure 3). The conclusion is that of the fossil fuels, only coal has staying power, in the sense of centuries of available supply. In a resource context, our fissionable resource base is at least comparable to our solid fossil resources. No global strategy which recognizes a potential climate problem can ignore the potential of the coal option.

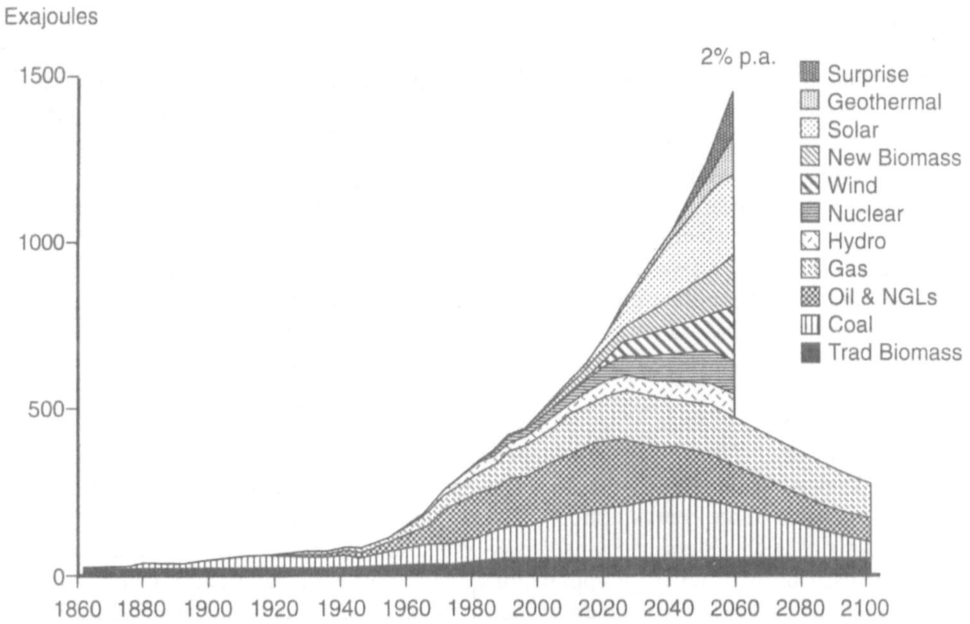

Figure 2. World Energy Use: Sustained growth 1860-2100. *Source:* Shell Group Planning PL/12.

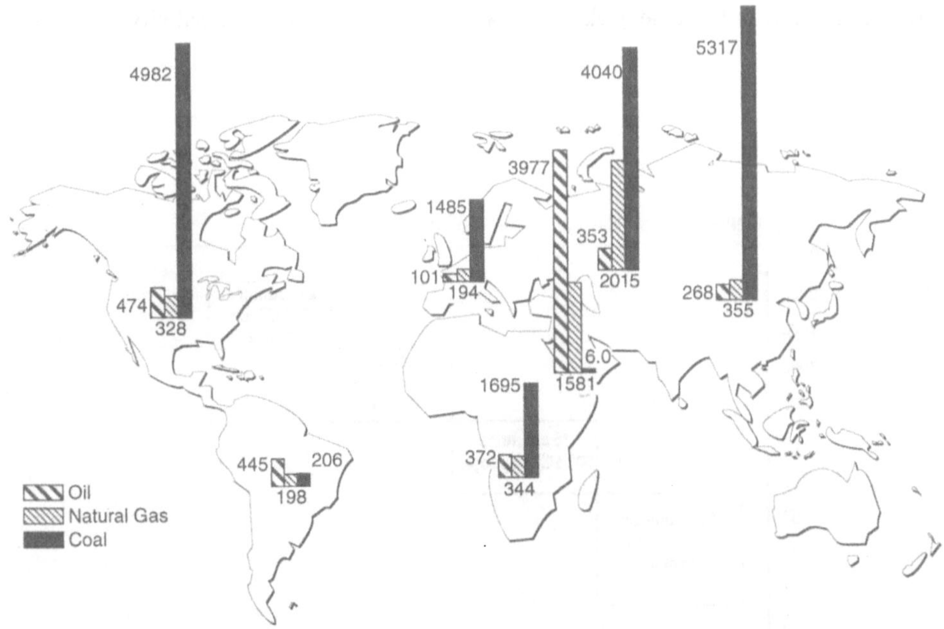

Figure 3. Global Energy Resources: Proven reserves in quads. *Source:* Energy Information Administration and the World Energy Council.

We must also add to this fossil and nuclear picture , the solar resource. It is essentially infinite and exploitable directly or through derivatives for energy, as well as other life sustaining purposes. Hydro, wind and biomass, while not new, are clean and substantial. Concerted efforts are being made to exploit the solar resource in the 21st century, and we agree with Shell that these efforts will pay off.

PROSPERITY AND ELECTRIC ENERGY

The common denominator for effectively using the big three -- coal, renewables and nuclear -- is through the medium of electricity. This has particular significance to the developing nations, since over half the world's population is now essentially without electricity, and most of it without any commercial energy services.

Electricity use has been very closely linked to economic prosperity and improvements in social welfare. But the electrification of the world has exaggerated the global disparities in energy use. Figure 4 shows a more than a 10-fold difference in electric generating capacity per capita between the rich and poor nations, correlated with a nearly a 100-fold difference in per capita GNP. Two points: First, we must work to close these gaps. I agree completely with the first priority of the World Energy Council, "to relieve the poverty in the developing world."

Second, the type and scale of technology appropriate for nations in the upper right hand box may be very different than those appropriate for the nations in the lower left hand box. Electrification, while contributing to a less energy intensive path, is also among the most capital intensive of the energy options. Thus it is imperative that strategies be crafted for each country that considers its human, energy and capital resources. Electrification should be focused on high leverage applications that improve education and health; and applications that will serve as the enablers in industrialization and economic development.

As the next century unfolds, the issue of global sustainability will begin to transcend the seaprate concerns of population, energy economy, health, social welfare, and the environment. New means for achieving sustainable development will be required, and efficiency is likely to form the backbone of all future strategies of sustainability.

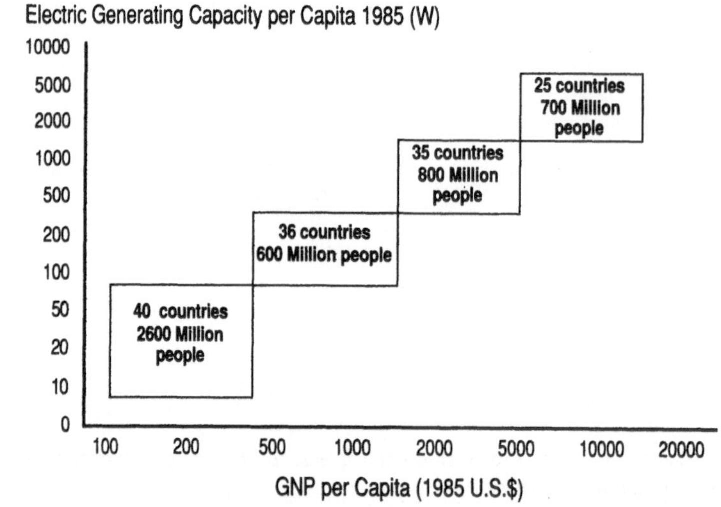

Figure 4. Prosperity and Electric Energy. *Source:* ASEA Brown Boveri.

ENERGY INTENSITY -- U.S. ENERGY/GNP RATIO AND ELECTRICITY FRACTION

It is illuminating to look at energy efficiency from an economic perspective. The U.S. went through a development cycle in which the energy input per unit of economic output climbed steadily from the beginning of the industrial revolution to just after World War I (Figure 5). At this point the energy intensity of the economy turned down. Why? Certainly because automobiles, furnaces, and industrial processes were becoming more efficient. But also in part because electricity was taking root in the economy. In the 1920s, the electric motor unit drive revolutionized manufacturing. And subsequently it revolutionized the office, home, and farm. As electricity continued to replace less efficient energy sources, the ratio of energy needed to produce a unit of GNP has steadily declined. Electricity's fraction of total energy is now approaching 40% in the U.S. and continues to climb.

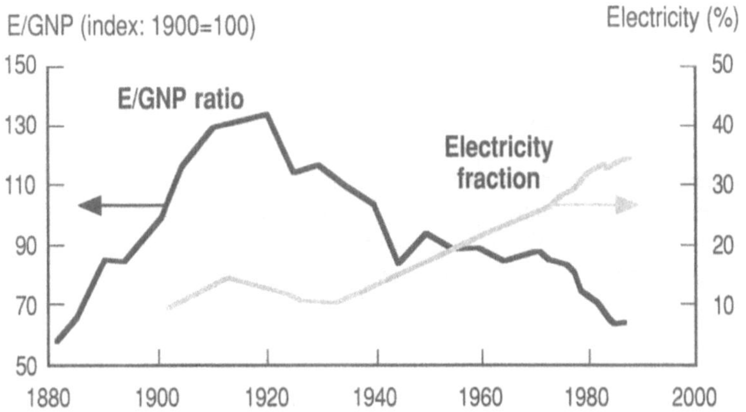

Figure 5. Energy Intensity: U.S. energy/GNP ration and electricity fraction (1880-1988). *Source:* Electricity in the American economy, Sam H. Schurr *et al.*, 1990.

ENERGY INTENSITIES OF INDUSTRIALIZED COUNTRIES AND DEVELOPING COUNTRIES

This phenomena is not unique to the U.S. Each of the industrialized countries has passed through a similar historic pattern in energy intensity, beginning here (Figure 6) with Great Britain in the 1880s and extending through Japan in the 1950s. Notice that over time, the peak of energy intensity for each country has progressively declined. This has enormous significance for the developing nations. Energy technology has advanced so rapidly in a historical context, that developing nations can pass through the same development cycle using only a fraction of the energy required a century ago.

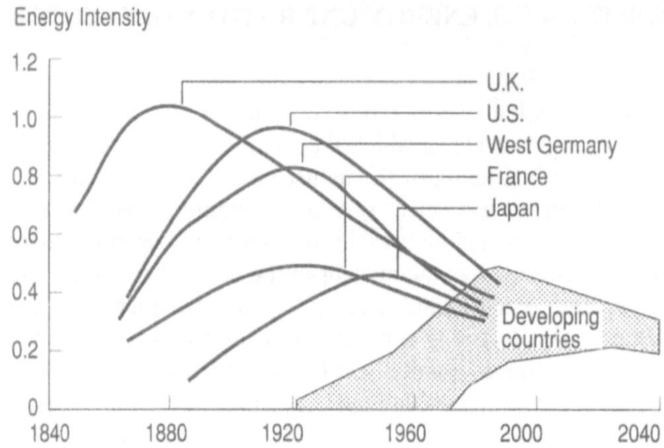

Figure 6. Energy Intensies of Industrialized and Developing Countries: Energy intensity can be reduced. *Source: Scientific American*, Sept. 1990.

END OF A PARADIGM

Given the importance of electricity in realizing economic growth and reduced energy intensity to date, what are the prospects for the future? Clearly environmental priorities now exist in our cities and at the global scale (Figure 7) that will influence future energy use patterns. Driven by the cumulative growth in U.S. environmental laws -- the Clean Air Act, he Clean Water Act, and RCRA -- to mention just a few, fossil power plant efficiencies have been declining in recent decades, reversing a hundred year trend of increasing efficiency and lower unit cost as economies of scale combined with better materials led to improved power plant efficiencies and economics.

EFFICIENCY TRENDS

But sustained technological improvement in combustion turbines and combined cycle concepts -- along with better oil and gas exploration and production technologies -- have created a new power plant paradigm (Figure 8).

A wide variety of advance combustion turbines are emerging that will eventually push combustion turbine efficiencies into the 60% range. The combustion turbine, burning natural gas, dominates new capacity additions in the U.S., and is readily transferable to any region of the world where gas or oil is available. The advantages include high efficiency, low-capital cost, low emissions, modularity, short lead time, and, at least for the foreseeable future, low fuel costs.

These machines can be used in simple cycle or coupled with a heat-recovery steam turbine, in a combined cycle mode. The technology has advanced rapidly so that machines with capacities in excess of 250 mW are now available in sizes that permit factory fabrication and railway delivery as the next slide shows.

TECHNICAL CHANGE VECTOR --GENERATION

The availability of low-cost, modular technology has spearheaded a fascinating shift in technical direction. We are moving (Figure 9) from one era dominated by the economy of

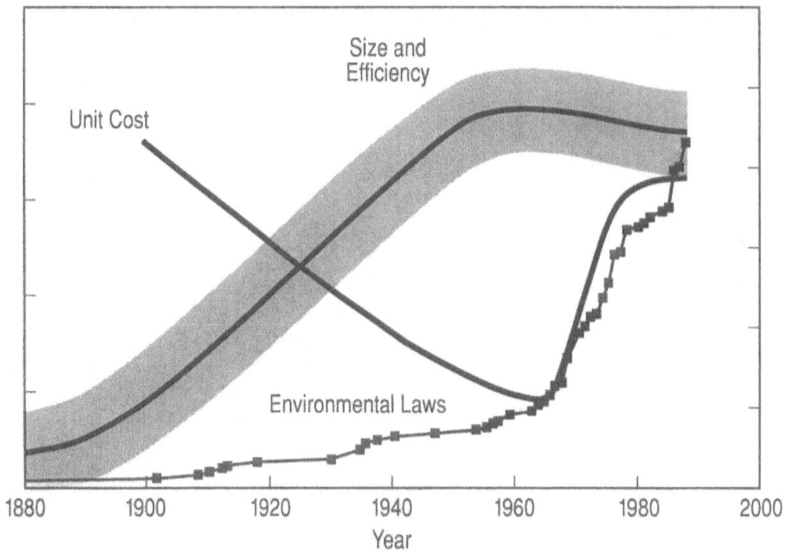

Figure 7. The End of a Paradigm?

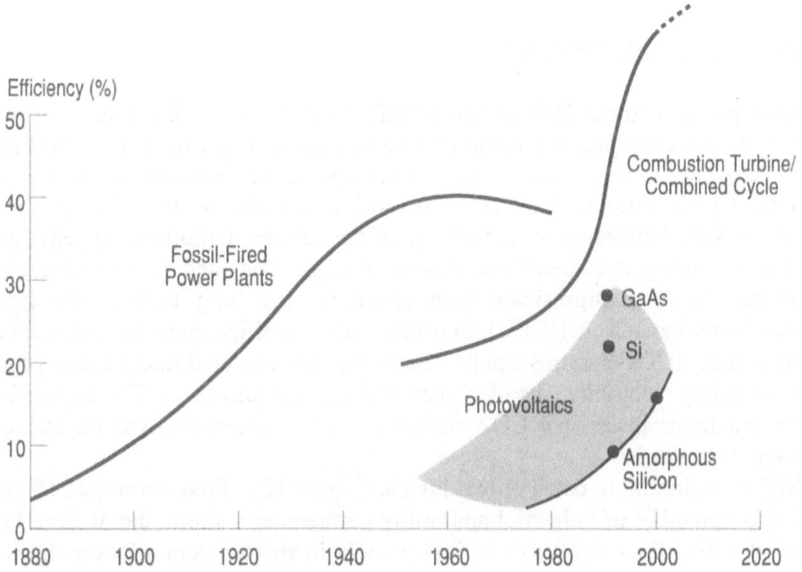

Figure 8. Efficiency Trends.

scale, to a new era marked by the "economy of precision." The precision of manufacturing replaces the laborious processes of on-site construction.

Combustion turbines and fuel cells rely upon gas. Gas, as seen over the sweep of the next century, is a transition fuel. As the price of natural gas rises, coal-derived gas will fill in.

Figure 9. **Technical Change Vectors Generation.**

IGGC GLOBAL DEPLOYMENT

Coal represents about 80% of the world's fossil reserves, a resource so large and geographically dispersed that the world dependency can only grow. Coal is destined to be the principle energy pathway for economic development in China and India over the next half century. By one estimate, China's growth will require one medium sized power plant to be built every week between now and the end of the decade. Unfortunately, environmental concerns are secondary to economic need in much of the Asian and developing markets.

Perhaps the most significant technology for the long term is the Integrated Gasification Combined Cycle (IGCC), in which coals is gasified, purified, and combusted in combined cycles. IGCC systems can be designed to convert coal into a broad portfolio of products, including electricity, chemical feedstocks, and liquid fuels. The high efficiencies of current machines lessen the CO2 burden per kWh generated, and far exceed other environmental standards.

IGCC is beginning to catch on worldwide (Figure 10). First developed 10 years ago by EPRI and a number of industrial and utility partners, at a site in the Mojave Desert of California, the 100 MW Cool Water facility brought in visitors from all over the world. It was the cleanest power plant in the world. Today, IGCC systems are being planned or under serious consideration in 21 countries. We believe this is just the start. Close observers of China's energy situation regard IGCC as China's best technology option for the long term.

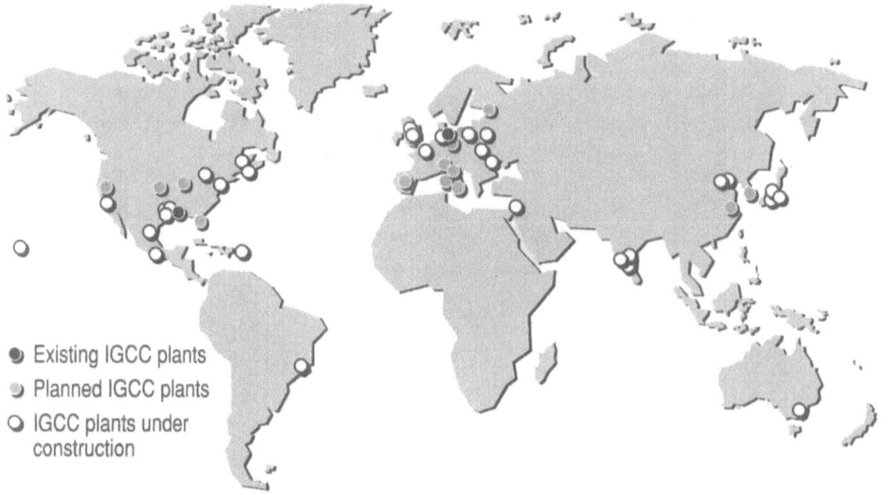

Figure 10. IGCC Global Deployment.

RENEWABLES

Renewables including hydro, solar photovoltaics, wind, and biomass (Figure 11) will contribute greatly to our energy mix in the next century. The resource base is simply too large and too dispersed to ignore, and technology is making great strides in improving efficiency and lowering cost. With renewables we need to think very long term.

Photovoltaic solar technology should head the list of promising technologies for the 21st century. The real long term promise of photovoltaics rests on the fact that it is part of the family of solid-statetechnologies which are still in the early, robust stages of technological discovery and development. Research continues to make great progress...cell efficiencies are climbing...manufacturing techniques are being adapted from integrated circuitry...and costs are falling. In the last decade, costs have fallen from about $1.00/kWh to about $0.30/kWh; and within 20 years we expect to reduce these costs to between $0.07 and $0.10 per kWh. When you can produce it on peak at the point of use for these kinds of costs -- with a resource that's free forever -- its time to take it seriously.

Photovoltaics is modular at scales appropriate for even low power applications in village life. India, for example, has 600,000 such villages which receive 12 hours of sunlight daily who could benefit from "pre-electrification" levels of power to run for example, simple lighting and cooking, a low power TV set, and in some households an efficient refrigerator.

Even though photovoltaics cannot compete today for bulk power generation, they are often the least expensive option for small, distributed applications in remote areas, even in

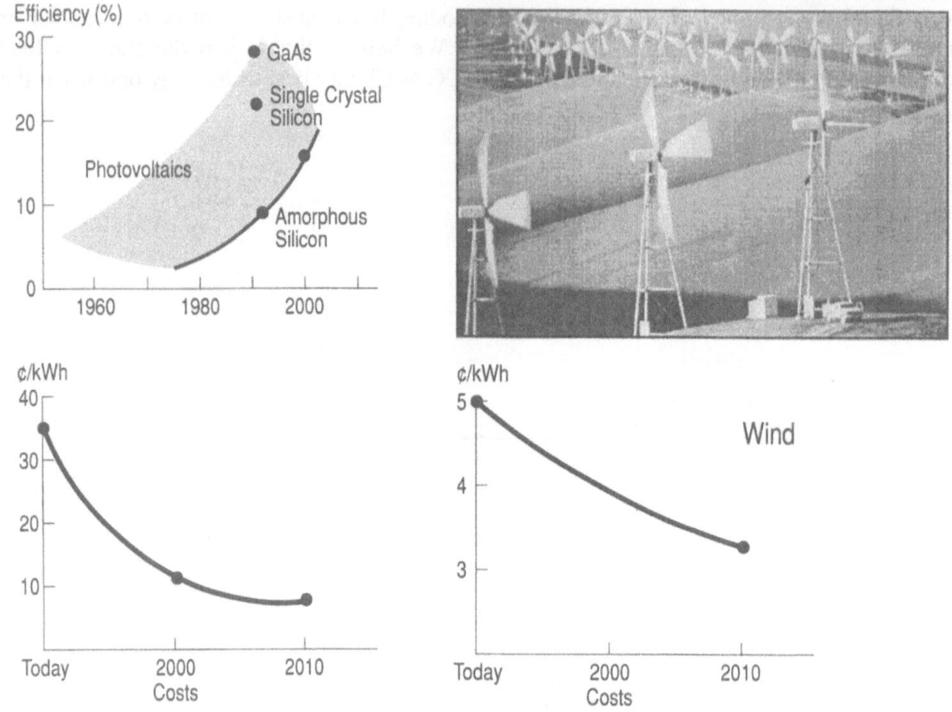

Figure 11. Renewables.

North America. The market is building as costs decline. Worldwide, photovoltaic sales are about 60 MW per year, and growing between 15-20% annually.

Wind

Wind energy is coming of age for power production. It has become competitive in the U.S....is rapidly taking hold in Europe...and is now poised for broad international deployment. There is increasing interest in the developing nations, particularly India.

Technologically, the leading edge in this field goes to the variable speed turbine developed by U.S. Windpower, EPRI and the DOE. It is capable of producing electricity for $0.05/kWh, given an average wind speed of 16 miles/hour. The turbines are rated at 350-450 kW, and can operate at wind speeds between 9 to 60 miles per hour. Fierce competition is setting in, and we should see a variety of new wind machines -- constant speed and variable speed -- emerge in the next decade. We expect to see wind power costs to drop to the $0.03-range within the next two decades.

NUCLEAR POWER PLANT PRODUCTION COSTS

In energy terms it seems imperative to me that we sustain the viability of our nuclear resource. The U.S. currently generates 20% (?) of its electricity from nuclear plants located primarily in the eastern two-thirds of the country. It competes with coal economically but complements it environmentally. Nuclear capital and operating costs vary significantly from plant to plant depending on age, type, and management practices. Figure 12 shows the range of running costs including fuel and O&M. If the variable cost can't compete with other types of generation it will increasingly difficult to sustain operation in the competitive markets ahead.

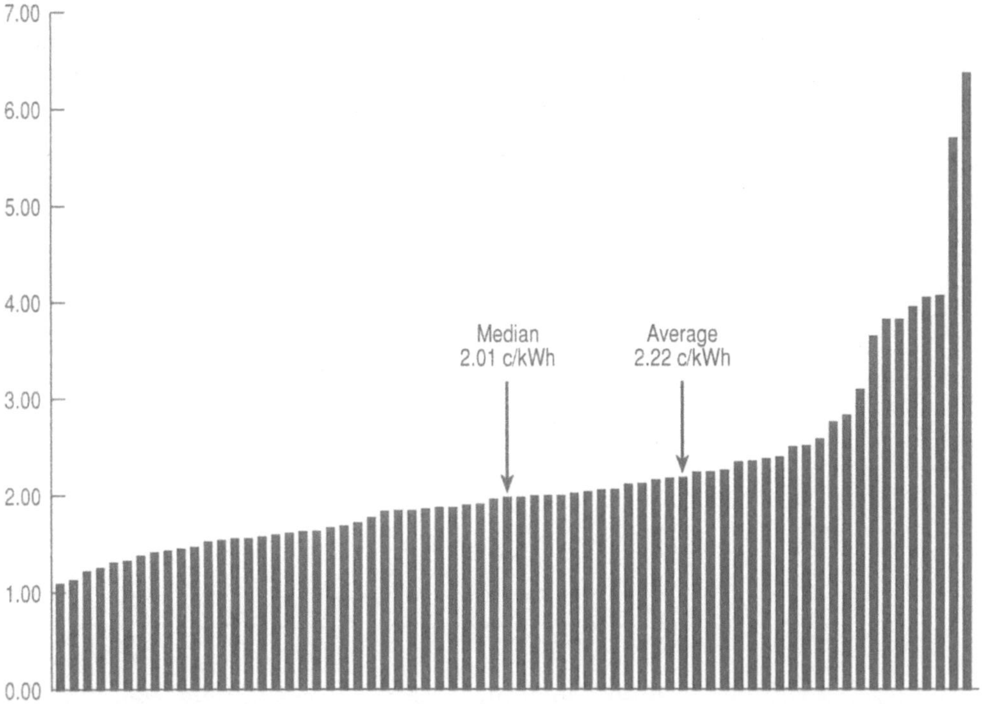

Figure 12. 1994 U.S. Nuclear Plant Production Costs (excludes plants with capacity factors <15%).

In the longer term, we must find political solutions to deal with spent fuel management, safety concerns, and the economic advantage that technology is providing other energy resources. Let me suggest, but leave for others to discuss here today, some possibilities. First in addition to ALWR technology, let's reconsider the potential safety and economic opportunities of high temperature gas cooled reactors. Gas turbine combined cycle efficiencies have far outstripped the steam cycle; perhaps nuclear could benefit here.

Secondly, spent fuel storage -- not disposal -- is what's needed. We should assume that future generations will need to utilize the fuel value that remains in today's spent fuel in tomorrow's breeder reactor. Let's manage the resource responsibly on an international scale until that time. And finally, we need to be prepared with a national solution to managing the consequences of competitive markets on our nuclear generation option. A common support infrastructure of critical size would be important in assuring the continued operation of marginal cost plants. Smilarly decommissioning costs could be kept to a minimum without compromising saftey if utilities didn't have to all go through the learning curve at the same time in case shutdown is required. Either higher gas prices, or rising greenhouse gas concerns, could shift the economics dramatically at which time mothballed units might make sense.

THIRD WAVE OF ELECTRIFICATION

Let me conclude with some remarks about trends in electricity consumption in the first half of the 21st century. This forecast (Figure 13) rests on the innovation potential inherent in electricity which makes it the energy form of choice at the point of use.

Electricity currently represents 36 percent of the energy resources consumed in the United States; its use has grown in two waves in electricity's first century. Many of us in this room have lived through the second wave and will recall that lights, radios and motors were the principle users of electricity in childhood. We've subsequently experienced a steady stream of new technology that has changed the home and workplace dramatically since 1940.

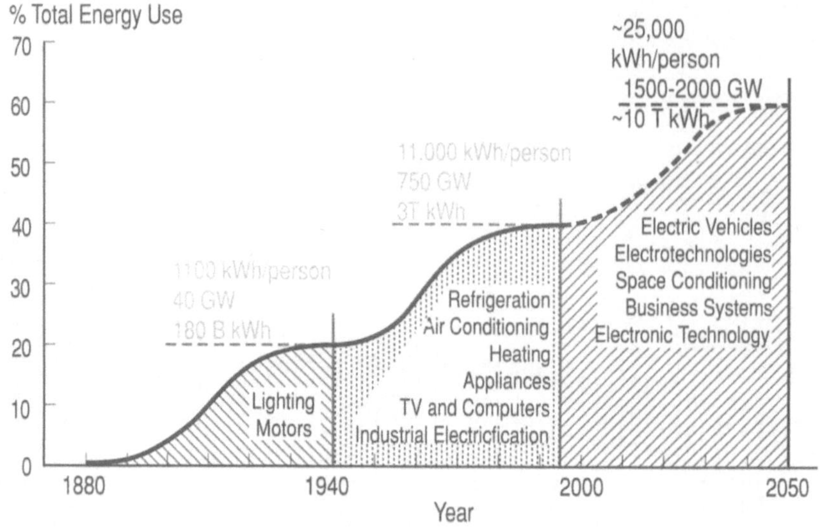

Figure 13. Promising Future.

From typewriters to word processors, carbon paper to Xerox, scrub boards to washing machines -- soon to be joined by microwave clothes dryers -- radio to TV, slide rules and the post office to computers globally networked, iceboxes to refrigerators, fans to air conditioners, blast furnaces to minimills, and on and on in industry, the office and the home. Energy consumption not only soared during this period, but the fraction used to provide electricity doubled. A grand and glorious first century! What lies ahead?

I'm a strong believer that we've only begun to exploit the power and versatility of electricity. Let's look at those possibilities that today's science and technologies suggest for the future, recognizing much of what we'll take for granted in 2050, we can't yet imagine. Again, total energy consumption will likely grow in the U.S., but likely at a slower rate as efficiencies increase and population growth rates stabilize. but electricity's fraction of primary energy consumption should continue to grow during this period with some suggested levels shown on the slide for 2050.

As suggested in this slide, electric vehicles, electrotechnologies and heat pumps will be key contributors to the next wave along with electronic business systems providing a broad spectrum of information, communication, education, commerce and recreation services. Electronic technology will permit miniaturization and portability in many product lines as battery technology grows in sophistication.

CONCLUSION

Finally, let me end with where I began. Can the environment tolerate the emissions burden that ten billion people will impose? That's a tougher question to answer definitively, but let me say it's not up to us (today's society) to solve <u>all</u> of the problems, a key one of which is population. "Innovation" is the most potent of our renewable resources and we should assume that we'll have a better understanding of and answers for the climate concerns, nuclear waste and the recycling opportunities that will extend our finite natural resources.

One thing that's evident is that the world will become smaller; we'll all be linked in real time and eventually by language as well as pictures. The world's aspirations will grow. Governments or their successors will be under increasing pressure to provide opportunity and access. Even historic religious and cultural barriers are likely to fall as TV and the global information network link us all.

We need to anticipate these aspirations by nations, institutions and individuals, and begin today preparing for the obligations we have to assure that the forces liberated by seeing and knowing can be peacefully channeled over the decades just ahead.

THE NEXT FIFTY YEARS OF THE PEACEFUL APPLICATIONS OF NUCLEAR ENERGY

Glenn T. Seaborg

Associate Director-at-Large
Lawrence Berkeley National Laboratory
Nuclear Science Division
1 Cyclotron Road
Berkeley CA 94720

I shall comment rather broadly and briefly on some diverse areas of the peaceful uses of nuclear energy and nuclear science.

I begin with a discussion of sources of energy for the future.

Environmental factors — the greenhouse effect, acid rain, and air pollution — are already constraining the use of fossil fuels and increasing their costs, and the impact of these factors is likely to grow in the future. The development and use of renewable energy sources, which represent varying ways of utilizing solar energy, are constrained by the technological and economic factors, as well as their limited availability in many regions, and this will keep their contribution limited. The sources of additional hydropower are limited.

The growth of nuclear fission power has been impeded by the perception of the need for safer operation, the obvious need for an adequate system for the disposal of radioactive wastes, and the imperative for the prevention of the proliferation of nuclear weapons.

In the United States, which has led the world in the birth and then the decline of the nuclear fission power industry, there are signs that its rebirth may be imminent. Presently, in the United States, in addition to programs to improve the present type of reactors whose normal operations have been environmentally benign, there is an aggressive program to develop new reactors with passive safety features. Passive safety features can be thought of as characteristics of a reactor which, without intervention of a human operator, will tend to shut a reactor down, keep it in a safe configuration, or prevent released radiation to the public. These features fall into two broad categories — features which are designed to prevent accidents from taking place, and those which mitigate the effects of potential accidents if they do happen.

In my opinion, the nuclear waste disposal problem is susceptible to technical solutions if the political obstacles can be overcome. The nuclear proliferation problem must be solved whether or not there is a growth of commercial nuclear fission power, and I believe this can be done through an increased safeguarding role of the International Atomic Energy Agency (IAEA).

Nuclear energy may be the key to further ventures into space during the next fifty years by human explorers, such as landings on the planet Mars.

Plutonium is the key to the long-term contribution of nuclear power in meeting the world's growing needs for energy. Breeder reactors based on plutonium-239 extract some

1 H																	2 He
3 Li	4 Be											5 B	6 C	7 N	8 O	9 F	10 Ne
11 Na	12 Mg											13 Al	14 Si	15 P	16 S	17 Cl	18 Ar
19 K	20 Ca	21 Sc	22 Ti	23 V	24 Cr	25 Mn	26 Fe	27 Co	28 Ni	29 Cu	30 Zn	31 Ga	32 Ge	33 As	34 Se	35 Br	36 Kr
37 Rb	38 Sr	39 Y	40 Zr	41 Nb	42 Mo	43 Tc	44 Ru	45 Rh	46 Pd	47 Ag	48 Cd	49 In	50 Sn	51 Sb	52 Te	53 I	54 Xe
55 Cs	56 Ba	57 La	72 Hf	73 Ta	74 W	75 Re	76 Os	77 Ir	78 Pt	79 Au	80 Hg	81 Tl	82 Pb	83 Bi	84 Po	85 At	86 Rn
87 Fr	88 Ra	89 Ac	104 Rf	105 Ha	106 Sg	107 Ns	108 Hs	109 Mt	110	111	(112)	(113)	(114)	(115)	(116)	(117)	(118)
(119)	(120)	(121)	(154)	(155)	(156)	(157)	(158)	(159)	(160)	(161)	(162)	(163)	(164)	(165)	(166)	(167)	(168)

LANTHANIDES	58 Ce	59 Pr	60 Nd	61 Pm	62 Sm	63 Eu	64 Gd	65 Tb	66 Dy	67 Ho	68 Er	69 Tm	70 Yb	71 Lu
ACTINIDES	90 Th	91 Pa	92 U	93 Np	94 Pu	95 Am	96 Cm	97 Bk	98 Cf	99 Es	100 Fm	101 Md	102 No	103 Lr
SUPER-ACTINIDES	(122)	(123)	(124)	(125)	(126)									(153)

Figure 1. Periodic table of the elements projected out to element 168. Atomic numbers of undiscovered elements are shown in parentheses. Included are undiscovered elements 122 through 153, which comprise the predicted and probably unattainable "superactinide" series.

50 to 100 times more energy from a given quantity of uranium than do current commercial reactors. Such reactors make it possible to utilize the large sources of much higher-priced uranium and thus can make nuclear energy the most abundant available energy resource and extend its availability indefinitely. It is unfortunate that the United States has stopped its program of development of the breeder reactor; I believe it will be necessary in the future to join the rest of the world in its development.

Now I turn to the role of nuclear fusion reactors, which in theory can have virtually unlimited sources of nuclear fuel. The present types can be developed for economic operation, if this should prove possible at all, on an unfortunately long time scale — that is, the magnetic confinement and the inertial confinement fusion reactors. Although such reactors do not have the fission product waste disposal problem attendant with fission reactors, they are accompanied by neutron emission leading to large quantities of radioactive by-products, and they are also plagued by the difficult problem of converting this neutronic energy into electric power. Required here are extremely high plasma temperatures or difficult laser implosion techniques requiring huge installations that have defied the ingenuity of investigators for many decades, with the prospects that such hurdles for economic power may not be overcome during the next fifty years, if ever.

Fortunately, there is another broad area of approach for the development and use of nuclear fusion that can, in principle, overcome these difficulties and be available on a much shorter time scale. Aneutronic fusion reactors utilizing colliding beams of reactants offer the possibility of much smaller size and economic construction and operation.

Another area for the beneficial applications of nuclear energy is in the use of radioactive isotopes in our space program, nuclear medicine, industry, agriculture, and the arts and humanities.

One of the uses of plutonium-238 is in the Space Nuclear Auxiliary Power (SNAP) units that have been used to power satellites or, more important, to power remote sensing instruments. SNAP sources (fueled by plutonium-238) served as power sources for instrument packages on the five Apollo missions, the Viking unmanned Mars lander, and the Pioneer and Voyager probes to Jupiter, Saturn, Uranus, Neptune, Pluto, and beyond, and will be the key to future more ambitious space projects during the next fifty years.

The use of radioactive isotopes in nuclear medicine contributes enormously to the prolongation of human life and the alleviation of suffering. Of the 30 million people who are hospitalized each year in the United States, one in three is treated with nuclear medicine. More than 10 million nuclear-medicine procedures are performed on patients and more than 100 million nuclear-medicine tests are performed each year in the United States alone. A comparable number of such procedures are performed in the rest of the world. The expected tremendous increase in the next fifty years, in expanding new types of diagnosis and treatment, is impossible to estimate.

The applications of radioisotopes in industry are numerous. Among these are: in manufacturing, for highly sensitive gauges to measure the thickness and density of numerous materials and to inspect finished goods for weakness and flaws; in the automobile industry, to test steel quality in cars; in the manufacture of cars, to obtain the proper thickness of tin and aluminum; in the aircraft industry, to check for flaws in jet engines; with construction crews, to gauge the density of road surfaces and subsurfaces; for pipeline companies, to test the strength of welds; for oil, gas, and mining companies, to map the contours of test wells and mine bores; and for cosmetic companies, to sterilize their products.

In addition, there are manifold uses in agriculture. In plant research, radiation is used to develop new plant types to speed up the process of developing superior agricultural products. Insect control is another important application; pest populations are drastically reduced, and in some cases, eliminated by exposing male insects to sterilizing doses of radiation. Fertilizer consumption is reduced through research with radioactive tracers. Radiation pellets are used in grain elevators to kill insects and rodents. Irradiation prolongs the shelf life of foods by destroying bacteria, viruses, and molds.

The useful application of radioisotopes extends to the arts and humanities. Neutron activation analysis is extremely useful in identifying the chemical elements present in coins, pottery, and other artifacts from the past. A tiny unnoticeable fleck of paint from an art treasure or a microscopic grain of pottery suffices to reveal its chemical makeup. Thus the works of famous painters can be "fingerprinted" so as to detect the work of forgers. Many old photographs thought to be beyond saving have been restored to a remarkable degree through the application of the neutron activation process.

There are almost unlimited possibilities for future research on transuranium elements, which already have reached up to element 111, an extension by 20% of the total of known chemical elements. When thinking in terms of the next fifty years, the tendency is to underestimate potential contributions.

Of central importance is a national policy discussion to retain and recover the potential stock of heavy transuranium nuclides now in storage basins at the Savannah River Plant in South Carolina. These are a priceless treasure and should not be discarded as irretrievable waste. This store can serve as starting material for the synthesis of increasing amounts of heavier transuranium nuclides for years to come.

Estimates suggest that 500 transuranium nuclides would have half-lives sufficiently long to be detectable experimentally (longer than a microsecond). The synthesis and identification of another half dozen or more elements seems likely; this would include the discovery of "superheavy elements" and the extension of the present peninsula of elements to connect with the "island of stability" which would be centered around element 114. Longer-lived isotopes than those now known will probably be found in the transactinide region especially among the early transactinide elements. It should be possible to study the chemical properties of elements beyond hahnium (element 105) and certainly of seaborgium (element 106).

The figure shows an imaginative periodic table extending all the way up to element 168. The completion of the actinide series, with the filling of the 5*f* electron subshell, occurs at element 103 (lawrencium). Elements 104 (rutherfordium) through undiscovered 112 are formed by filling the 6*d* electron subshell, which makes them homologues in chemical properties with the elements hafnium (Z = 72) through mercury (Z = 80). Elements 113 through 118 would result from filling of the 7*p* subshell and are expected to be similar to the elements thallium (Z = 81) through radon (Z = 86). The 8*s* subshell should fill at elements 119 and 120, thus making these an alkali and alkaline earth metal, respectively. Next should come the filling of the 5*g* and 6*f* subshells, 32 places in all, which I have termed the "superactinide" elements, followed by the filling of the 7*d* subshell (elements 154 through 162) and 8*p* subshell (elements 163 through 168).

Although we can feel confident that this is the approximate form the periodic table should assume, we, unfortunately, will not be able to verify much of this experimentally because the half-lives of the nuclei are too short and there are no nuclear synthesis reactions available on earth to reach such heavy elements. However, I believe it will be possible to add some six new known elements (perhaps somewhat more) to our periodic table.

The art of one-atom-at-a-time chemistry will advance far beyond what can be imagined today to make it possible to study the chemistry of heavier and heavier elements. All of this will result in the delineation of relativistic effects on the chemical properties of these very heavy elements, which might thus be substantially different than those expected by simple extrapolation from their lighter homologs in the periodic table.

Such a research program will require, for success, the availability of apparatus and equipment of increasing complexity, versatility, and power. Central will be the need for higher neutron flux reactors for sustained operation as a research tool and to produce large quantities of transplutonium nuclides for use in the research and as target materials as a source of the presently known and expected nuclides. Higher intensity heavy ion accelerators must be built and the means of coping with the heat generated in the target by such intense beams must be developed in order to overcome limitations due to small nuclear reaction cross sections. Increases by orders of magnitude in heavy ion intensity should make possible nuclear synthesis reactions with secondary (radioactive) beams of neutron-excess projectiles, which might greatly increase the range and yields of sought-after new nuclides. Improved methods for handling safely and efficiently and making chemical measurements on increasing quantities of the highly radioactive transcurium nuclides must also be developed.

These are some of the expanded peaceful applications of nuclear energy and nuclear science that can be expected in the future. We can be certain that any attempt to make predictions on a fifty year time scale is doomed to failure. The rate of progress today suggests that we will in general underestimate the magnitude of progress during such a long time.

Chapter II

Energy Supply Projections of Primary Fuels, Renewables, Conservation, and Nuclear Power

THE ROLE OF NATURAL GAS IN ELECTRIC POWER GENERATION IN THE TWENTY-FIRST CENTURY

Steven I. Freedman

Gas Research Institute
Chicago, Illinois 60631

INTRODUCTION

Industrial society is energy, capital and information intensive. As a nation makes the transition to an industrial society, the degree of electrification in terms of the electricity share of point-of-use energy consumption increases as does the electric intensity of GDP and the electricity consumption per capita. A substantial number of lesser developed nations are rapidly expanding their economies and consequently their power generation capacities. As these developing economies expand, the environmental consequences of their primary energy choice will have profound effects due to the implicit long term commitment to specific technologies. In the industrialized world we are realizing the extent to which technology choices have extremely large inertia's. Historically, coal was the low cost energy resource because of the ease of discovery, simplicity of mining and the availability of low cost labor in societies undergoing industrialization. While coal is less expensive than natural gas on an energy basis, the recent emergence of low cost high efficiency gas turbine combined cycle equipment, combined with increased discoveries of modest cost gas worldwide, results in natural gas combined cycle produced electricity being the lowest cost power in many locales, besides being by far the cleanest of the fuel fired alternatives.

Four factors of prime importance bear on the issue of power generation primary energy; resource availability, economics, environmental impact, and equipment technology. In three of the four categories natural gas has clear advantages; in the other, resource availability, a distortion of the perception of the natural gas resource in the United States has misled us in the past and we need to increase our knowledge of the geophysical occurrence of natural gas in order to create the greatest economic benefits in the 21st century. In the world as a whole, natural gas' share of the primary energy market has been increasing[1] to the point where gas is now a major fuel for power generation in a substantial portion of the world. Natural gas use for electric power generation has two main questions; how much gas will be used and in what manner will it be used.

Economics and Politics of Energy
Edited by Kursunoglu *et al.*, Plenum Press, New York, 1996

The need to eventually use non-fossil energy sources is apparent for two reasons: eventually we will deplete the readily accessible (economically recoverable) fossil resource base, and at some time in the future, the conversion of additional large quantities of naturally sequested carbon (coal, oil, and gas) to atmospheric carbon dioxide could result in unprecedented rapid global climate change. At present, there are technical, economic, practicality, safety and political concerns with the principal forms of non-fossil energy resources for bulk power generation, nuclear and solar. In the interim, before an economically attractive and politically acceptable form of non-fossil bulk power generation emerges, the world, and the United States in particular, will have to rely on fossil fuel for the majority of our power generation primary energy. This narrows the competition to coal, oil, and natural gas fuels. Besides the shortage of domestic oil at a competitive price and the economic and political consequences of major energy imports, oil is more expensive than coal as a fuel for steam (only) power plants and is neither lower cost nor as clean as gas for combined cycle power plant use. The fossil fueled choices then become natural gas combined cycle, coal/steam and coal/gasification combined cycle. Natural gas and gasified coal can generate power with comparable atmospheric emissions, NOx, SO_2 and particulates as the coal-derived fuel gas is cleaned with processes mostly based on the same technology as is used to clean natural gas obtained from natural subsurface reservoirs. However, coal gasification plants have additional solid wastes. Coal/steam plants, when compared to either natural gas or coal gasification combined cycle plants, emit greater amounts of NOx, particulates and SO_2, and considerably more CO_2 and hazardous air toxics per unit of electrical energy generated. It is reasonable to expect that in the early part of the twenty-first century a level playing field regarding the permissible emissions from fossil fuel-fired powerplants will have been established and that new gas- and coal-fueled plants will have to meet a common set of emission requirements per unit of electrical energy generated. In the competition between coal-gasification and natural gas combined cycle power plants with comparable emissions of NOx, SO_2 and particulates, natural gas plants emit half of the CO_2 of the coal-gasification version, are much lower in cost and, based on current forecasts of gas prices, produce power at total cost (fuel, capital recovery, interest and O&M) appreciably less than the coal-gasification alternative.

POWER GENERATION ALTERNATIVES

Investment decisions are made on the basis of the expectation of future profit, with consideration of various risks including that of changing future fuel price. Where does natural gas power generation economics stand today? What roles will it play in the 21st Century? Combined cycle power plants are over 50% (HHV, 55% LHV) efficient today with plants of up to 54%/60% being offered commercially and future increases in efficiency being expected. These plants are the cleanest commercial fossil fuel fired central station powerplants, emitting essentially zero SO2, particulates, and solid waste, and the lowest NOx and CO_2. They are also the lowest price, \$600/kW, of all bulk power baseload alternatives. Natural gas fuel is sufficiently low in cost, around \$2.00 to \$2.25/MMBtu delivered to most utilities and large industrial power generators in the United States such that the majority of economically selected new power plants use natural gas combined cycle equipment. Costs of powerplants have been decreasing in the last few years as utilities, independent power producers, manufacturers, and architect/engineers are becoming much more competitive for the market-based electricity supply business of the future. The best data on the capital costs (including interest during construction) efficiencies of baseload natural gas and coal fired powerplants are summarized in Table 1. Presented in Figure 1 is the sum of the capital charge, fuel and O&M costs for the three alternatives. Natural gas costs are presented for two, economic instances, today's ten year levelized cost for delivery to powerplants, and a higher value typical of ten year levelized prices in the near future. In all cases, the natural gas combined cycle cost is the lowest. As expected, the cost of electricity from coal and nuclear plants are highly sensitive to interest rates. Non-fuel operation and maintenance expenses strongly favor natural gas combined cycle plants as they require smaller staffs to operate, have essentially zero waste disposal cost and have much less boiler tube wastage and component replacement costs when compared to coal combustion and gasification plants, and nuclear plants.

In addition to bulk power generation, natural gas use is increasing in industrial cogeneration. The combination of high efficiency from forthcoming small (1 to 10 Mw) advanced gas turbines along with their low emissions and modest cost may permit large scale siting of modular generation and distributed cogeneration units serving individual commercial and industrial consumers and being employed by utilities as part of their distribution systems at the substation level. This later market has been envisioned as being served by fuel cells in urban areas with severe ozone problems. The use of clean high efficiency gas turbines and reciprocating engines is expected to expand the market for small power sources broadly as the economic and environmental advantages of the high fuel use efficiency of cogeneration and the electric grid support value of distributed generation become more widely realized.

Small cogeneration plants were used a century ago when electricity was the high technology energy form still in its technical and commercial infancy. Due to the needs of the technology of the nineteenth century, coal fired steam boilers and reciprocating steam engines, and the economics of scale of coal delivery, ash removal, boilers and engines central station utility power generation became the dominant source of electricity for the twentieth century. What has changed in the last decade is the increased reliability and computer (or microprocessor) control of small engines, turbines, and fuel cells. With such modern products, small scale remotely dispatched power generation has become practical. The combination of gas price deregulation and the low wellhead price of natural gas has resulted in attractive economics of industrial cogeneration using natural gas. Inasmuch as the gas distribution system in many areas can deliver additional energy without major capital investment in infrastructure, gas fueled distributed generation may offer lower cost to electric utilities faced with the high cost of expansion of their electric distribution system, especially in crowded urban areas. In addition to gas playing a major role in new central station generation, gas can be the energy form of choice to deliver electricity to substations and the distribution infrastructure with near point-of-use modular energy conversion. The term "modular generation" encompasses both customer (on site) generation and grid connected distributed generation. It can be either fuel cell, reciprocating engine or gas turbine technology. Fuel cells hold the promise of ultra-clean (1 ppm NOx operation), attractive efficiency 36% to 55%, and extremely quiet generation. The challenge to fuel cells becoming a widely used commercial product is in the ability of the fuel cell developers to manufacture and deliver their product at a competitive capital cost.

Reciprocating engines and gas turbines are price competitive in the 3 to 5 Mw size range, with engines being favored in smaller sizes and turbines in larger sizes. While engine exhaust has higher NOx levels than those of the newest gas turbines, stoichometrically operated engines have been cleaned up with catalytic converters to comparable levels at modest cost. The battle for modular generation will be fought over equipment availability, cost, and emissions. When designed and built for long term, high reliability service, such as in pipeline compressor stations, natural gas engines have

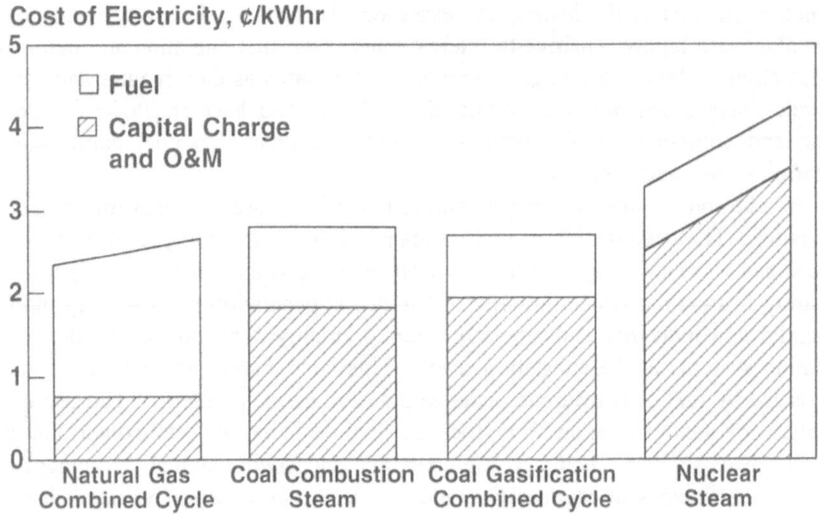

Figure 1. Comparison of Cost of Electricity in 2000.

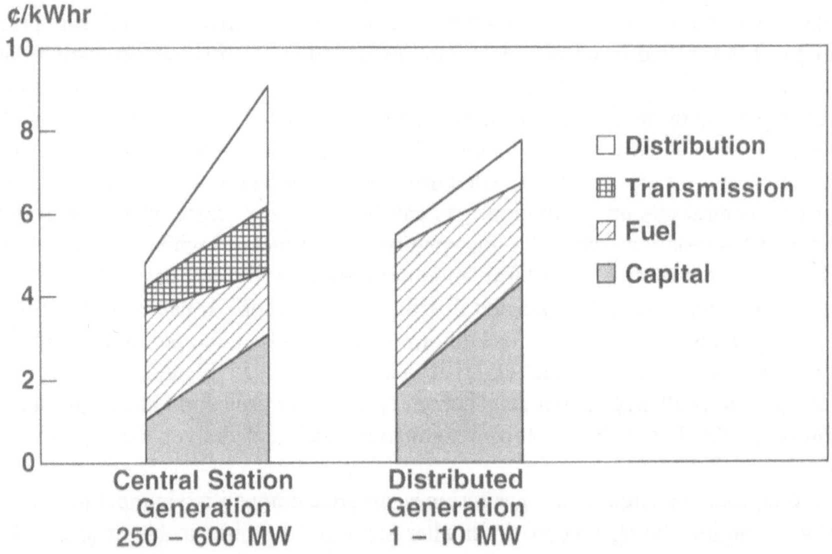

Figure 2. Representative Breakdown of Cost of Electricity.

Table 1. Economic Comparison of Baseload* Powerplants in 2000[9,10]

Cost Element	Natural Gas Combined Cycle		Coal Combustion Steam	Coal Gasification Combined Cycle	Nuclear LWR
Capital Cost, $/kW	600	600	1500	1600	2000
Efficiency (HHV)	54%	54%	35%	47%	35%
Capital Charge @ 9%, ¢/kwhr	0.68	0.68	1.69	1.80	2.25
Fuel, ¢/kwhr					
Natural Gas @ $2.50/MCF	1.60				
Natural Gas @ $3.00/MCF		1.90			
Coal @ $24/Ton			0.98	0.73	
Uranium Fuel and O&M[11]					1.00 to 2.00
O&M (non-fuel)					
Natural Gas Combined Cycle	0.05	0.05			
Coal Combustion, Steam			0.10		
Coal Gasification Combined Cycle				0.15	
Total Cost of Electricity	**2.33**	**2.63**	**2,77**	**2.68**	**3.25 to 4.25**

*Baseload assumed 8000 hrs/year

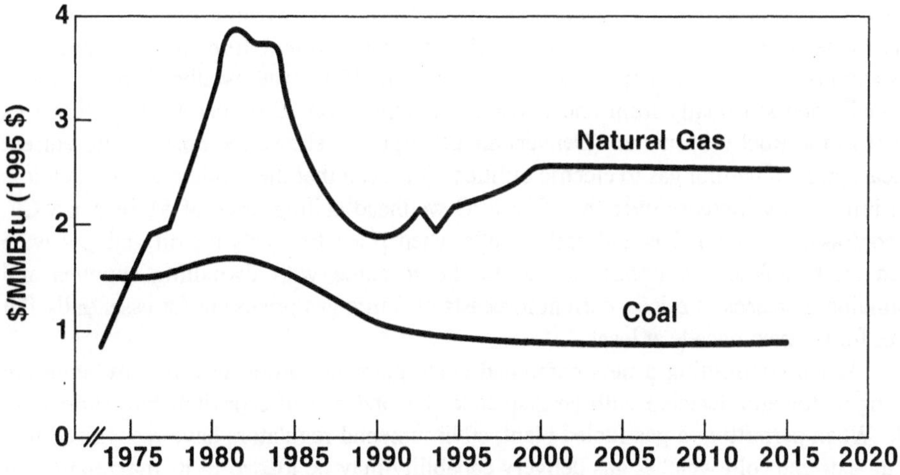

Figure 3. Fuel Price to Electric Utilities.

delivered the high availability needed for power generation service, albeit at a cost higher than their light duty cousins. While many question the reliability of engines based on their personal experiences with low cost gasoline engines intended for modest life service, industrial engines designed to meet customer prescribed durability have proved highly reliable in year-round continuous duty service. Proponents of modular generation base its attractiveness on cost and efficiency with the implication that it will be available when needed, thereby minimizing electric demand service costs. In order to succeed in the marketplace, the technology must offer advantageous economies, environmental acceptability, and reliable service in a single product. Figure 2 shows how the various elements of the cost of electricity to the user; capital, fuel, transmission, and distribution add up such that for some customers on-site, or near-site generation is economically advantageous.

Based on current and near term fuel cost forecasts, natural gas combined cycle plants are economically and environmentally superior to coal-fueled alternatives for bulk power generation, and natural gas fueled modular generation is expected to supply substantial amounts of electricity in the future. Key questions are: how long will the price of natural gas continue to be low enough so that gas based electricity will command an economic advantage over coal, and when will non-fossil fuel based power generation become economic?

GAS SUPPLY

Let us now examine the natural gas supply picture in the United States. In the United States, natural gas was "cheap and abundant" in the 1950s when the transcontinental pipelines were being built. The Philips Petroleum-Federal Power Commission (FPC) Supreme Court case of 1954 upheld the authority of the FPC to regulate the (wellhead) price of natural gas purchased by pipelines for sale in interstate commerce. Such regulatory action discouraged purposeful exploration for natural gas with the consequential shortages of the 1970s. This artificial shortage of natural gas disappeared when gas was decontrolled in the late 1970s and 1980s. Gas decontrol resulted in adequate incentive for exploration for natural gas such that the well head price in the mid-to-late 1980s dropped to below the 1972 regulated price, in spite of inflation reducing the value of the dollar to less than half in the same time period. Technology advances in exploration and production, especially the use of high powered computer tools to determine the precise structure of deep underground formations, contributed substantially to this accomplishment. Today, in 1995, the wellhead price varies regionally and seasonally from under \$1.00 to slightly over \$2.00 per MCF, with prices lowest in the Rocky Mountain (Denver) area.[2] Figure 3 shows the historic and current forecast price of natural gas to electric utilities. It is seen that these prices are a much less than inflationary increase over the 1972 FPC wellhead ceiling price of \$1.38 per MCF. As contrasted to the 1970s and early 1980s when price forecasts for oil and gas were based on beliefs of an imminent end to the resource with dwindling supplies and continuing increases in prices, current forecasts of future gas prices are for essentially flat prices for the next decade at least.

When committing a new combined cycle plant to natural gas, the investors are making a strategic decision with consequences beyond that of a (switchable) short term fuel. When permitting a gas fueled plant, environmental regulators have recognized that during extreme cold weather gas delivery capability may be loaded to its maximum, and consequently are receptive to permitting operation on a backup fuel, such as distillate oil, for a reasonable extent of operation. When such backup fuel use is included in the economic analysis of gas use, the annual average (not 365 days/year) gas cost and deliverability becomes quite attractive. While gas prices suffered an anomalous temporary peak in the mid-1970s to early-1980s period, due to consequences of regulation, gas prices otherwise have been level in terms of real dollars and are expected to continue to be stable into the foreseeable future. Competition involving a substantial number of energy supply companies, with extremely large cash flows, provides confidence that competitors will apply resources to ensure that the customers of natural gas will not be troubled with supply interruptions and price fluctuations.

As a consequence of gas deregulation, futures markets have developed for natural gas. The best known gas futures market is the "Henry Hub", named for a gas pipeline interchange point in Texas. Options on gas price futures are routinely traded in large quantities on the New York Mercantile Exchange (NYMEX), and reported in the Wall Street Journal, for deliveries in up to 18 months. Future options are readily available for up to ten years from major investment firms, and contracts for longer periods have been made. Complicated "Swaps" are often involved. Recent data [3] of Henry Hub futures shows that gas options for annual use at prices of up to $2.00/MMBtu are available through the year 2003 and that even by the year 2007 annual gas prices can be covered by options at a gas price below $2.10. These option prices differ from actual gas cost by the present value of the option cost reflected into the gas cost at the time of delivery. The complexities of gas futures trading have been made practicable by the recent advances in personal computers and communications equipment; another way that technology advances permeate into improvements in economic efficiency in the energy sector.

How much gas do we have; that is the trillion dollar question. A principal historical view of gas resources has been one of finite reservoirs and a known time to depletion. In the field of natural gas, this concept has proven limiting for four reasons. Firstly, exploration technology has improved through the use of more powerful computer processing of seismic data. Secondly, production has improved through more efficient and deeper drilling. Thirdly, economic technology has been developed for operation in deeper offshore waters. And lastly, and most importantly, exploration has occurred in heretofore unimaginable areas such as the North Sea, the mountains of Columbia and various frontier regions. Natural gas is now being commercially produced domestically from coal beds, tight sands and shales.[4] How much better seismic and computer technology will be in the future, how much deeper and further offshore economic production operation will be, and in what new locales gas will be found are questions which cannot be answered with the type of precision normally practiced in economic planning.

The key unknown in quantifying the extent of gas use for power generators in the 21st Century is the extent to which supplies will be available at a competitive (for power generation) price. The concept of resource depletion has been important at least for the last two hundred years since Malthus' alarmist view that population growth would rise faster than food production. While the concept of an upper bound on natural resources rests on the awareness that our planet is finite and consequently limits exist for material resources, previous attempts to quantify such bounds have almost always been found to be grossly pessimistic when reviewed a generation (25 years) later. The recent review by

Hodges[5] states that "Contrary to predictions from the 1950s through the mid-1980s, persistent shortages of non-fuel minerals have not occurred, despite prodigious consumptions, and world reserves have increased." "World reserves of gold were estimated at one billion troy ounces in 1971, and after more than 20 years of rising production, world reserves now amount to more than 1.35 billion troy ounces." In a similar manner, "the estimated 7 billion barrels of natural petroleum remaining available in the ground in the United States" stated authoritatively by David White in 1920 have been produced a substantial number of times[6]. In the case of natural gas, M. King Hubbert's encompassing article on energy resources[7] stated that as of 1970 "for the 48 states" "the ultimate amount of natural gas would be about 1075 trillion cubic feet. (TCF)" Since then, about 500 TCF have been produced and the current study by the Potential Gas Committee[8] reports that current estimates of the gas resource base varies between 930 TCF (based on current exploration and production technology) to 1758 TCF (based on advanced exploration and production technology). This 1995 report should be compared with the first report of the Potential Gas Agency in 1977[7] when similar estimates from 17 sources for the period from 1968 to 1974 were compared, showing that on top of the 12 year proven reserve base, there were between 400 TCF and 2250 TCF of additional estimates of future potential supply beyond 1974.

Perhaps one of the most revealing indicators of the differences between the studies summarized in 1977 and 1995 is the extent of water depth in the Gulf of Mexico deemed practical to drill in; in the 1970s, it was 300 to 1000 feet and in 1995, it was 1000 meters. Similarly in the last ten years, advances in computer hardware and commercially available software have substantially increased our knowledge of the exact location of deep geologic subsurface structures likely to contain gas. These, and other technical advances in gas (and oil) exploration and production technology, along with the entrepreneurial risk taking inherent in drilling in locales previously considered dry by others (such as Kuwait until 1937), support the view that "it is too early to close the patent office" on new discoveries. How much more gas there is in the United States beyond the 1000+ TCF-50 year supply is not known. Not known is not the same as known to be zero, even though zero is a possibility.

Planning for the future must include the concept of a horizon beyond which additional information, discoveries, inventions and new ideas will penetrate the market. On the basis of information currently available, we may conclude that there is a major role for natural gas to play in power generation in the next fifty years and that forecasts should be periodically updated to account for developments in technology, resource assessment, environmental considerations, and other factors which often evolve in unforeseen directions.

REFERENCES

[1]International Energy Annual 1993, DOE/EIA-0219 (93), U. S. Department of Energy, Washington, May 1995

[2]Holtberg, Paul D., et. al., Baseline Projection Data Book; 1996, GRI, Washington

[3]Bennett, Porter, Natural Gas Supply & Delivery: Long Term Contracts at Competitive Prices are Available, Blast Furance Gas Injection Workshop, Gas Research Institute, October 19, 1995

[4]Woods, Thomas J., Technologies and Improved Strategies Have Lowered Finding Costs and Opened New Exploration Opportunities, Gas Issues and Trends, Gas Research Institute, Washington, August 1995

[5]Hodges, Carroll Ann, Mineral Resources, Environmental Issues and Land Use, Science, Vol 268, Page 1305, June 2, 1995
[6]Pratt, Wallace E., Toward a Philosophy of Oil-Finding, Bulleting of American Association of Petroleum Geologists, Page 2231, December 1952
[7]Hubbert, M. King, The Energy Resources of Earth, Scientific American, Page 31, September 1971
[8]Hanley, John M., Comparison of Estimates of Ultimately Recoverable Natural Gas in the United States Gas Resource Studies No. 1., Colorado School of Mines, April 1977
[9]Todd, Douglas M., and Joiner, James R., IGCC Cost Study, EPRI Thirteenth Conference on Gasification Power Plants, San Francisco, October 19 - 21, 1994
[10]Simbeck, Dale, Air-Blown Versus Oxygen-Blown Gasification, Gasification: An Alternative to Natural Gas, Institution of Chemical Engineers, London, November 22, 1995
[11]International Conference on Economics and Politics of Energy, Miami, Beach, November 27 - 29, 1995

CONTRASTING NUCLEAR GROWTH POTENTIAL: CENTRAL/EASTERN EUROPE VERSUS SOUTHEAST ASIA

Raymond J. Sero

General Manager
International and Major Projects
Westinghouse Energy Systems

In my position as general manager of international and major projects for Westinghouse Energy Systems, I spend 50% of my time in Central/Eastern Europe and the other 50% in Southeast Asia. After a discussion with the organizers of this conference, I agreed to discuss my impressions of the potential for new nuclear additions in each of these global markets.

I have chosen to do that by looking at these two unique markets and contrasting and comparing them. I will examine the political/social factors which influence energy decisions in these markets, and try to forecast how those factors might ultimately affect future nuclear construction.

Table 1. The Regions

Central/Eastern Europe	Southeast Asia
Bulgaria	India
Czech Republic	Indonesia
Finland	Japan
Hungary	North Korea
Romania	PRC
Russia	Philippines
Slovakia	South Korea
Slovenia/Croatia	Taiwan
Ukraine	Thailand

I will begin by clearly defining the two markets I'm considering. The Central/Eastern Europe market consists of those non-European-union countries, who presently have nuclear power plants as part of their energy production mix. With one exception, these countries were members of COMECON. The combined population in this market is approximately 275 million people, with a predicted GDP growth over the next three years of 3.5-4.3% per year. The Southeast Asian market consists primarily of the new market tigers. Three of these countries - Indonesia, North Korea, and Thailand - do not have nuclear programs today, but they have announced their intentions to do so. I've added one country — the Philippines — just to round-out the market and, maybe, for sentimental reasons. The combined population in this market is 2.75 billion people with a predicted GDP growth rate over the next three years of 7.3-8.2%.

At this point, considering two known facts - that the population of the one market is an order of magnitude greater than the other, and that the GDP growth rate of that same market is expected to be double that of the smaller market - one would have to ask why bother with any other analysis? I believe, however, some other interesting aspects of these markets are worth discussing.

MARKET SIMILARITIES

First, these diverse markets share some important characteristics which could affect future energy decisions significantly.

Table 2. Similarities in Eastern/Central Europe and Southeast Asia Markets

Economies in transition • Dramatic changes = dramatic problems
Magnitude of new investments • Where will the money come from? • Where will it go first?
Significant environmental concerns
Institutional challenges • Privitization of industry • Pricing based on cost • Regulatory framework

Both markets consist of economics which are in a state of transition. The Southeast Asian market countries are experiencing growth booms and changes in their political structures and philosophies. Most of these countries have standards of living below those of developed markets in North America and Western Europe. These countries want improved infrastructure, which means improved living conditions. In Central/Eastern Europe, the situation is similar. These countries are all experiencing the pain of transition from centrally planned economies to market-driven economies. For the last five years, these economies have experienced decline, and the standard-of-living for the people in these countries has worsened.

These countries all want to reverse this recent past. They all have industries which can be managed to produce economic growth, and some have started. They all want better living conditions, which again requires an improved infrastructure. A single difference, of course, is the perception that while adequate energy resources exist in these countries, energy efficiency is poor. In either case, the need for improved living standards creates a significant tension in these markets to implement policies to obtain those improvements, and the developed countries must recognize this tension and the need to satisfy the expectations of the people. Failure to do so could cause momentous political instability.

A second important aspect of the two markets is the magnitude of the investments which will be needed to meet infrastructure and economic expectations. It will take significant capital to improve the conditions in these markets, and only a small portion of that capital will be self-generated.

The third significant common aspect of these markets is the need to deal with environmental issues. In Central/Eastern Europe, present environmental damage that has been commonplace in their industries must be dealt with. So, despite the perceived abundance of energy in this market, I would argue that new energy additions are mandatory to alter the ongoing environmental damage. In Southeast Asia, the environmental issues are more recent in emerging, but just as severe. Anyone who has visited the forests of Bohemia or experienced the traffic in Beijing/Seoul/Bangkok understands the need for future environmental focus. Of course, these environmental concerns have a positive impact on the likelihood of nuclear as the selected alternative for energy additions.

Finally, the two markets have much in common with respect to the institutional challenges that must be addressed in conjunction with energy additions. In both of these markets, there is a shifting from government owned/controlled energy production to "privatized" ownership. Hungary is accepting proposals for private ownership of their fossil fleet. In Taiwan and the Philippines, IPP projects are being promoted. In addition to privatization, these markets all feature an interesting fact - power pricing to the consumer is not based on the cost of generation. This creates a dichotomy - how does a private investor become interested in a business opportunity in which it cannot control its rate-of-return? Finally, and most important to the nuclear option, is the fact that only in Japan and possibly South Korea is there a strong regulatory legal framework within which nuclear plants can be constructed and operated with an understanding of nuclear risk and liability protection.

MARKET DIFFERENCES

Now that I have discussed why these markets are similar and why nuclear is attractive in these markets, let's look at the differences in the regions which will influence nuclear decisions.

The first area was addressed previously, with just one added note: while GDP growth for the C/E Europe countries is predicted to be greater than 3.5%, the reality of the present is that these countries - with the exception of Finland and the Czech Republic - have continued to experience slow or no growth. This type of market does not attract potential investors.

On the positive side for C/E Europe, new additions which are planned are not based on "starting from scratch." New nuclear units were under construction throughout the market when what we knew as the "cold war" came to an end. Hence, many plants have already been begun. At Belene in Bulgaria, 100% of one unit's mechanical equipment has been delivered and is in storage in warehouses on the site. At Mochovoce in the Slovak Republic, one unit is more than 85% complete. Of course, this is not the condition in Southeast Asia where new nuclear construction must start from a green field - with the exception, of course, of the Philippines.

One very important difference between these markets is the absence of a "role model" or leader in the C/E Europe market from which the smaller countries can learn and follow tried practices. MINATOM in Russia is certainly a huge industrial giant, employing over one million people in various commercial and defense-oriented nuclear activities. But it is a poor model for developing and adapting a complete, integrated nuclear program. In Southeast Asia, on the other hand, the Japanese and South Korean programs are major successes.

A difference between the markets - surprising, at least to me - continues to be the indifference in the Central/Eastern European market toward the need for a strong safety culture. I'm not talking about having a 5,000 man nuclear regulatory body or volumes of regulations. I'm talking about attitudes of the personnel. I have been working in this area of the world for over five years, and I have seen change, but I believe more is needed, especially if outside capital is to be made available to support new nuclear construction. On the contrary, the governments within Southeast Asia appear to be more focused on the need for a strong safety culture and a legal and commercial framework in which the nuclear option can be successful. This culture issue goes beyond the operators, of course, and extends to every facet of the legal framework within the particular countries. An understanding of the nuclear liability issue and the implementation of the necessary laws to support a strong nuclear industry are also required.

So far, most of these differences tend to favor the Southeast Asia market. One strong difference, however, which sheds favorable light on the C/E Europe market is the fact that its long-term political stability is a top-priority issue with the G7 countries. A "relapse" in the politics and philosophies of countries like Russia and Ukraine is not a favorable outcome for the stability of the region. As a result, I believe that the political climate promotes G7 decisions to encourage and support programs in the C/E Europe market which would lead to stable infrastructure and improved economic conditions. In this regard, I believe that financing can and will be made available to support infrastructure development. Such financing could take the form of grants, soft loans, and/or expert credit agency support. Such a policy change will favor construction in this vital market.

A final note on differences concerns the fact that none of the Central/Eastern European economies have ever experienced market pricing for energy. In the case of the Southeast Asia market, some of the countries, in fact, have cost-based pricing which makes them more attractive to the outside investor.

Table 3. Differences in the Regions

Central/Eastern Europe	Southeast Asia
Stagnant economies	Growing economies
Nuclear investments "sunk"	"New" nuclear construction
No "leading" force	Successful Japanese and Korean programs
Weak nuclear safety culture	Strong nuclear safety culture
G7 desire to achieve political stability	Higher level of political stability
Central planning/pricing	Mixed pricing policies

CAPITAL REQUIREMENTS

I mentioned that I chose these countries because they either had nuclear programs or intended to have them. I also mentioned the issue of the capital needed to finance the programs. On this table, I took the most recent established and published nuclear plans for each country and tried to estimate the capital requirements.

Basically, I used $2B per GWE to estimate the cost for the forecasted construction. This is probably a little low and results in a underestimate for the Southeast Asian market. It could be somewhat high, however, for the C/E Europe market because a number of those new additions have already been started, and many of them are more than 60% complete. Nonetheless, it is worth noting that the capital requirements in Southeast Asia are probably more than three times those in C/E Europe. Not only are these numbers large, $200B (+) over the next eight years or so, but also these numbers are a fraction of the capital forecasts that have been made for all energy additions in these markets. New gas and steam plants are planned in each of these markets, with capital requirements which are an order of magnitude greater than these nuclear numbers.

Table 4. Present Plans for New Nuclear Additions

Central/Eastern Europe		Southeast Asia	
Bulgaria	2 GWe by 2005	India	8 GWe by 2005
Czech Republic	2 GWe by 1998	Indonesia	2 GWe by 2005
Finland	2 GWe by 2010	Japan	26 GWe by 2005
Hungary		North Korea	2 GWe by 2003
Romania	2 GWe by 2005	PRC	18 GWe by 2010
Russia	12 GWe by 2005	Philippines	
Slovakia	2 GWe by 2010	South Korea	10 GWe by 2006
Slovenia/Croatia		Taiwan	5-6 GWe by 2006
Ukraine	6 GWe by 2004	Thailand	
Capital Requirements	$45-60B		$140-160B

FUNDING SOURCES

Where will the money come from? The obvious places, of course. As Table 5 illustrates, most of the capital will be generated through self-financing and Export Credit Agency loans.

Some will be through capital markets, however, and in this regard the Southeast Asia market has a distinct advantage.

Table 5. Primary Sources of Funds for Future Expansion

Central/Eastern Europe	Southeast Asia
G7 grants	
Self-financing	Self-financing
Export credit agencies' support	Export credit agencies' support
	Capital markets
Commercial loans (limited)	Commercial loans
Private investment - Equity funds - Individual investor	Private investment - Equity funds - Individual investor - Public offerings

The Czech Republic, and to a large extent Finland could use the alternative of capital markets today. Most of the other C/E European countries, however, do not currently have that luxury. As was mentioned before, Russia and Ukraine have access to G7 grants as part of any large-scale political stabilization process.

POTENTIAL NEW NUCLEAR ADDITIONS

With all of this discussion, it's obvious that the question of new nuclear additions is highly complex. I've given you some of the commonalities between the markets, and I've discussed the differences which will have an impact on the nuclear decision. Finally, before pursuing our forecasting, let's look at "generic" factors which also influence any ultimate nuclear decision. In Table 6 I've listed the present "generic" factors which influence a nuclear decision. In that regard, I have also indicated the relative importance of the factor – is it positive, is it high?

Table 6. Political/Social Factors Influencing New Nuclear Additions

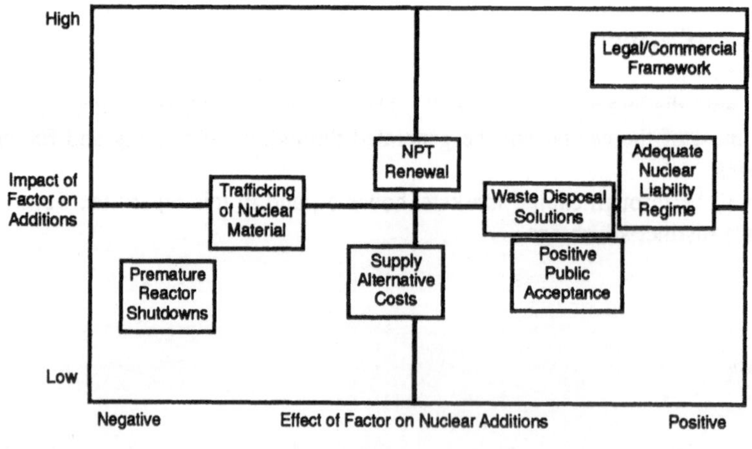

First and foremost is the question of the legal/commercial framework within any country which supports any energy option. Closely aligned with this question is the nuclear-specific issue of nuclear liability. Do the country's laws and commercial policies permit third-party ownership, do its laws support an attractive return on investment, do its laws protect the investor and contractor, and can a legitimate rate-of-return be calculated?

Waste disposal is critical in any country in the throes of a decision-making process. If the country recognizes the issue and provides an integrated solution to it, the likelihood of a positive nuclear decision is increased. Just such an approach is being pursued by the Czechs. Their laws will cover all aspects of the nuclear option and, when adopted, will provide the framework for a positive nuclear future - much like that of Japan. Obviously, it is mandatory for a commercial nuclear future that the Nuclear Proliferation Treaty be renewed in some fashion. (This topic was discussed in another part of the conference.) That issue is acknowledged at the center of the chart.

The most critical economic issue which challenges the nuclear option is simple: is it cost-effective? In many countries, cost-effective alternatives are available and these will be pursued as opposed to nuclear additions.

Finally, two generic issues of minor importance today appear on the left side of the chart. They concern the recent acknowledgement of nuclear material trafficking and the premature shutdown of operating reactors in the U.S. I believe these issues will increase in importance in the global decision-making process in the future.

THE FORECAST

I've contrasted the two markets and discussed my views of the political and social factors which influence those markets. Now, it's time to do what I promised - forecast the nuclear potential in these markets.

Recognize, this is not a simple matter, and to embark on such an endeavor smacks of foolishness. In fact, forecasting any kind of reality for energy additions is a "black art." And nuclear forecasting is even more of a devil's game. But, I'll plunge forward, nevertheless.

In creating the forecasting model shown in Table 7, I tried to limit the factors which would have an impact on my forecast. I chose three fundamental elements for the analysis.

- The need for any electricity addition at all. This is of great importance.
- The desirability of pursuing the nuclear option from the decision-maker's point of view.
- And, the ability to build a nuclear plant given the conditions and infrastructure in the country.

The chief contributors to the question of desirability, in my analysis, are:

- The total capital required - smaller is better.
- The nature of public reaction - with an optimal score equated to no significant public response/negativity.

Table 7. Forecasting Model

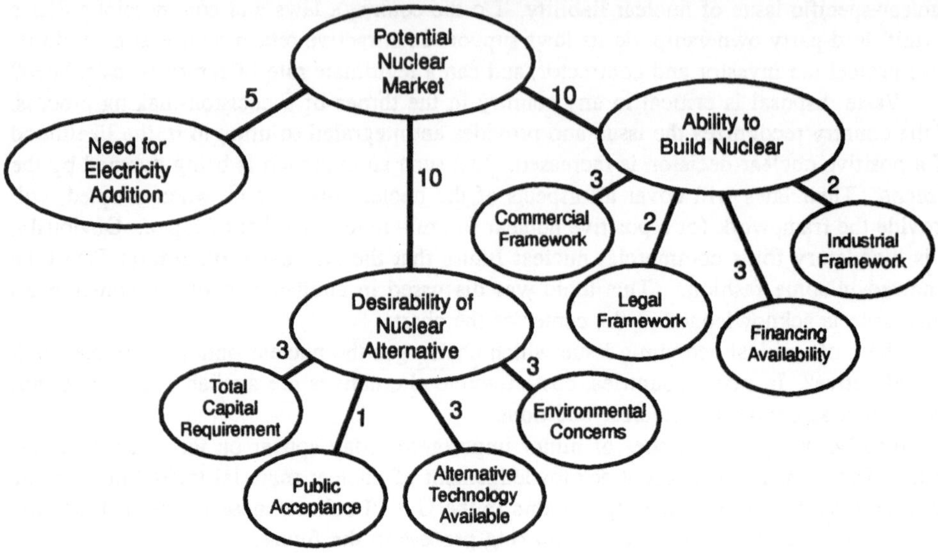

- The availability of energy alternatives – where no feasible economic alternative produces a maximum score.

- A measure of whether or not the environmental concerns in the country are real. If they are real and they are to be attacked, a high score is obtained.

The chief contributors to the question of ability to pursue the nuclear option are:

- The present commercial infrastructure – if it supports significant investment and provides for investor protection and returns, a high score is obtained.

- The legal framework supporting nuclear construction should exist with nuclear liability protection, insurance, ownership laws for assets, etc. — if a legal framework exists, a higher score is obtained.

- The local financing capability is critical — if self-financing is available, a maximum score is obtained.

- Finally, an area not addressed so far is the availability of a strong industrial infrastructure to support nuclear construction. This is extremely important with regard to the ultimate cost. If a strong industrial infrastructure exists, a higher score is obtained.

Based on those factors, what are the results?

Table 8. Forecasting Results

	Electricity Need	Total Capital	Public Acceptance	Alternative	Environmental Concerns	Commercial Framework	Legal Framework	Industrial Framework	Financial Availability	Total	Potential for Addition
Bulgaria	5	1	1	3	3	1	2	2	3	21	
Czech Republic	4	3	1	3	3	2	2	2	3	23	
Finland	3	1	0	1	1	2	2	2	2	14	
Hungary	2	1	0	2	3	2	1	2	1	14	
Romania	4	1	1	2	2	1	1	1	2	15	
Russia	2	2	1	0	1	1	1	2	1	11	
Slovakia	3	3	1	3	3	1	1	1	2	18	
Slovenia/Croatia	4	1	1	3	2	1	1	2	2	17	
Ukraine	4	3	1	3	3	1	1	2	2	20	
India	4	1	1	3	3	2	2	2	3	21	
Indonesia	3	1	0	2	3	1	1	0	2	13	
Japan	4	1	0	3	3	3	2	2	3	21	
North Korea	3	1	1	2	1	0	0	0	3	11	
PRC	5	1	1	3	3	3	1	2	3	22	
Philippines	5	3	0	1	2	3	3	0	2	19	
South Korea	4	1	1	2	3	3	2	2	3	21	
Taiwan	5	1	0	3	3	3	2	2	3	22	
Thailand	3	1	0	2	3	1	1	0	2	13	

Following my model, the assignment of importance and assessment of the present circumstances, the following observations can be made:

- The Czech Republic, Bulgaria, and Ukraine are the best bets not only to conclude present nuclear construction, but to go forward as well. The Czech Republic will probably not only complete Temelin, but will also wind up supporting Slovakia in completion of the Mochovoce units. The Ukraine is a sleeping giant with great needs, few alternatives, and great political significance.

- The Russian program probably will not materialize. The availability of alternatives, the willingness of investors to pursue those alternatives, the low regard for environmental issues, and the indifference toward establishing a conducive legal/commercial framework to encourage construction will hold Russia back.

- The nuclear programs of Japan, the PRC, Taiwan, and South Korea will continue into the future barring any unseen political issues.

- The construction program in India will continue, despite the U.S. sanctions.

- Finally, there is no good reason for a North Korean program to exist - except that a political decision motivates it.

- Oddly enough, considering their history, the Philippines really ought to consider a future program.

WORLD ENERGY USE - TRENDS IN DEMAND

Randy Hudson

Engineering Technology Division
Oak Ridge National Laboratory
Oak Ridge, TN 37831-8038

INTRODUCTION

In order to provide adequate energy supplies in the future, trends in energy demand must be evaluated and projections of future demand developed. World energy use is far from static, and an understanding of the demand changes underway in various regions of the world will assist in planning for and meeting those energy needs. This paper evaluates global energy use by both primary energy type and region over a thirty year period from 1980 through 2010. It is hoped that such an analysis will provide a useful perspective for those concerned with the economics and politics of energy. Owing to the old adage that a picture is worth a thousand words (and to the large number of figures), the narrative in this paper will be kept to a minimum; the message is contained in the figures themselves.

The data presented in this paper come from the U.S. Department of Energy's Energy Information Administration (EIA).[1,2] The EIA is an independent statistical and analytical agency within the Department of Energy. As such, EIA collects and publishes historical and projection data as a service to energy managers and analysts, both in government and in the private sector. The historical data presented here have been largely derived from published sources and from reports from U.S. Embassy personnel in foreign posts. The projections of world energy demand are derived from EIA's World Energy Projection System (WEPS).[3]

THE BIG PICTURE

It is useful to first consider total world energy consumption. As shown in Figure 1, total consumption has been growing at approximately 1.6 percent per annum since 1980. The EIA reference case projects growth to continue at 1.8 percent per annum. However, any projection into the future is not without uncertainty. EIA addresses this uncertainty by developing sensitivity ranges for each projection. The sensitivity ranges in Figure 1 represent a 1.1 and 2.4 percent per annum growth rate and are based on deterministically varying the assumed economic growth rate and energy intensity (i.e., energy consumed per dollar of GDP) in the WEPS model. A further discussion of the sensitivity ranges can be found in Reference 2. In order to enhance the clarity of the figures, the sensitivity ranges associated with the subsequent projections will be omitted from the plots. The reader should bear in mind, however, that each projection has a range of uncertainty which is documented in Reference 2.

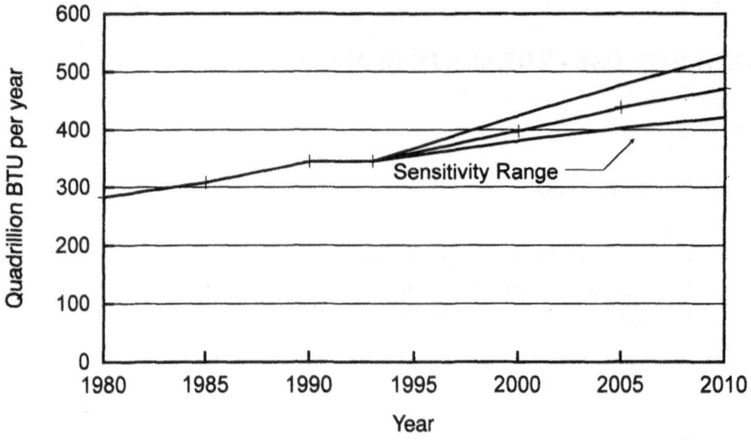

Figure 1. Total world energy consumption per annum.

Which primary fuels are used to satisfy this demand? As shown in Figure 2, oil is clearly the most heavily consumed resource. Further, fossil fuels provide over 85 percent of the world's energy.

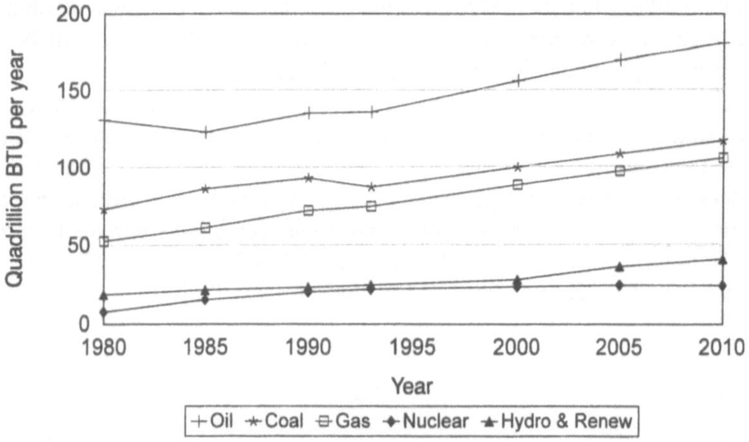

Figure 2. World annual energy consumption by fuel type.

Which regions of the world are the primary consumers of the energy shown in Figure 1? While historically North America and Europe have been the predominate consumers of energy, the Far East and Oceania region (e.g., Australia, China, India, Japan, Korea, Taiwan, etc.) is experiencing a tremendous growth in energy demand. It is currently projected that the Far East will have a growth rate in energy consumption that is three times that of North America. Figure 3 shows that by 2005, it is projected that the Far East and Oceania, with its 3.5 percent per annum growth rate, will surpass North America in annual energy consumption and become the largest consumer of energy by region.

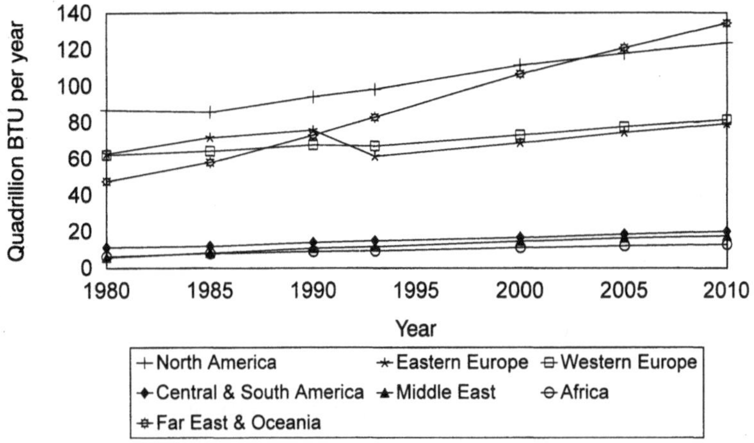

Figure 3. World energy consumption by region.

ANALYSIS BY ENERGY SOURCE

Turning our attention to the consumption of individual fuels, Figure 4 shows the global consumption of oil by region. Consistent with the earlier observation, the Far East is becoming a predominate consumer of petroleum, having surpassed Western European consumption in the last five years. By 2010, it is thought that the Far East's oil consumption will rival that of North America.

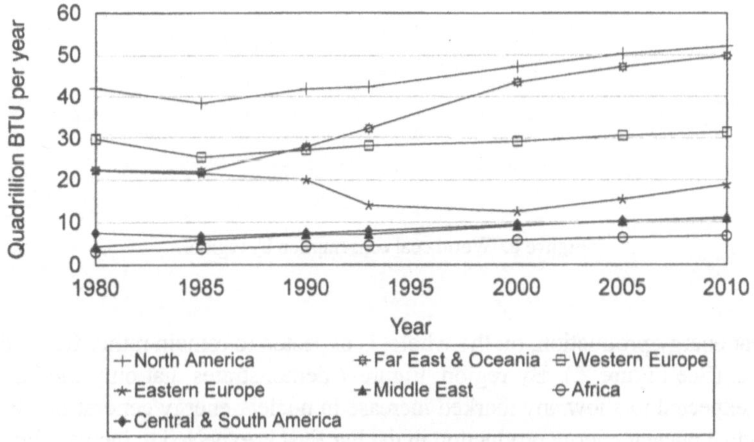

Figure 4. World oil consumption by region.

Reflecting the use of indigenous resources, Figure 5 shows that natural gas consumption is highest in Eastern Europe and North America. Owing to the relative difficulty in transporting natural gas, it is not expected to be a major near-term fuel source in satisfying the growing Far East energy demand.

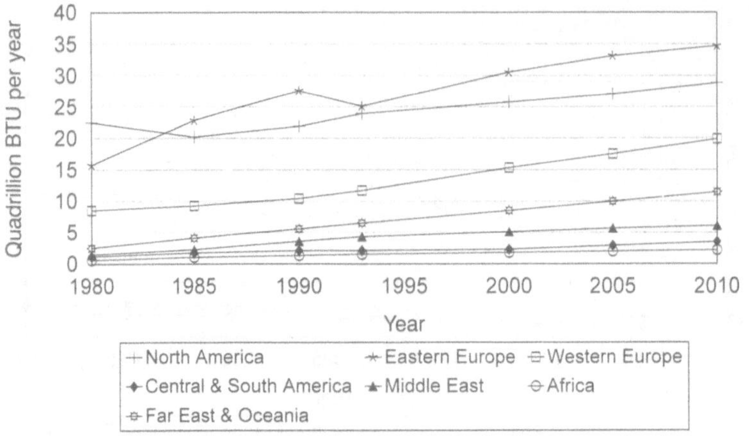

Figure 5. World gas consumption by region.

Tied to the high growth in energy demand, the consumption of coal in the Far East and Oceania is expected to grow at a 3.7 percent per annum rate; tripling the annual consumption between 1980 and 2010 (Figure 6). China, alone, is responsible for 70 percent of the Far East region's projected annual coal consumption in 2010. Given the environmental impact of coal burning, other regions' use of coal is expected to stabilize or decrease.

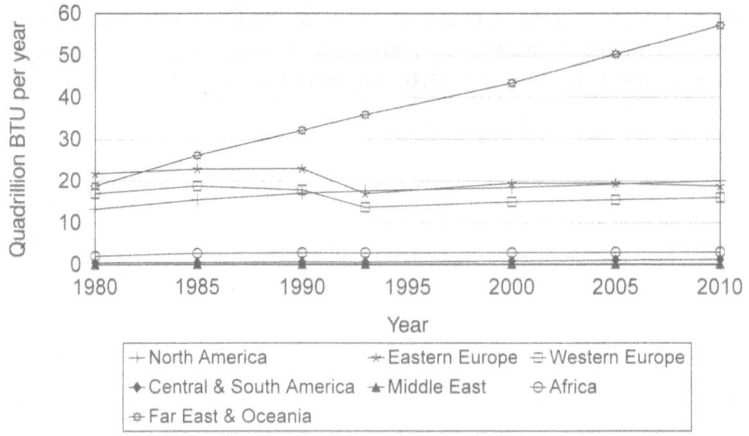

Figure 6. World coal consumption by region.

Nuclear energy production, on the whole, is expected to remain rather flat in the 1990 - 2010 period. (See Figure 2.) By region, Figure 7 demonstrates that only the Far East and Oceania is expected to show any marked increase in nuclear energy generation. From 1993 to 2010, annual nuclear energy production in the Far East is expected to increase by 2.3 quads (2.3 x 10^{15} Btu). At a typical plant heat rate of 10,000 Btu/kWh and an assumed capacity factor of 75 percent, this corresponds to 30 gigawatts of new nuclear electrical capacity.

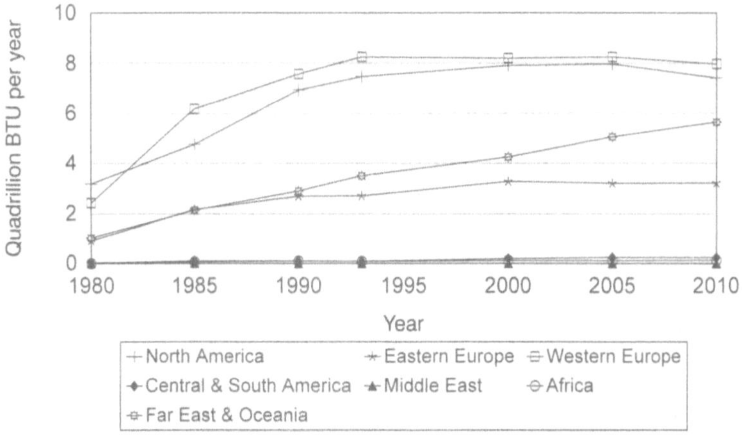

Figure 7. World nuclear energy consumption by region.

Finally, consumption of hydro and other renewable energy sources is shown in Figure 8. In this case, EIA is projecting a large increase in the use of this energy resource for North America and the Far East. In North America, annual consumption of hydro and renewable energy is expected to increase 8 quads between 1993 and 2010, which corresponds to approximately 120 gigawatts of capacity, if used for electricity generation. Similarly, an increase of over 5 quads is projected for the Far East.

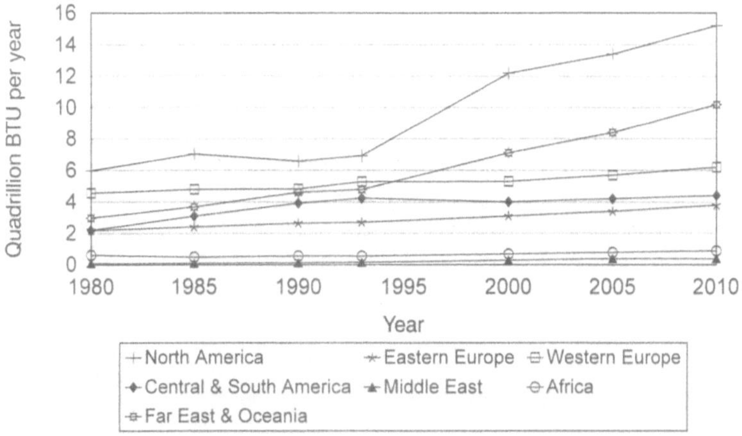

Figure 8. World hydro and renewable energy consumption by region.

ANALYSIS BY REGION

As shown earlier in Figure 3, four out of seven regions consume the vast majority (90 percent in 1993) of world energy: North America, Far East and Oceania, Western Europe, and Eastern Europe. As the energy mix is not the same for these regions, this section will provide more details on consumption patterns for each of these four regions.

In North America, oil is the most heavily consumed energy resource. As shown in Figure 9, oil consumption exceeds both the use of natural gas and coal combined.

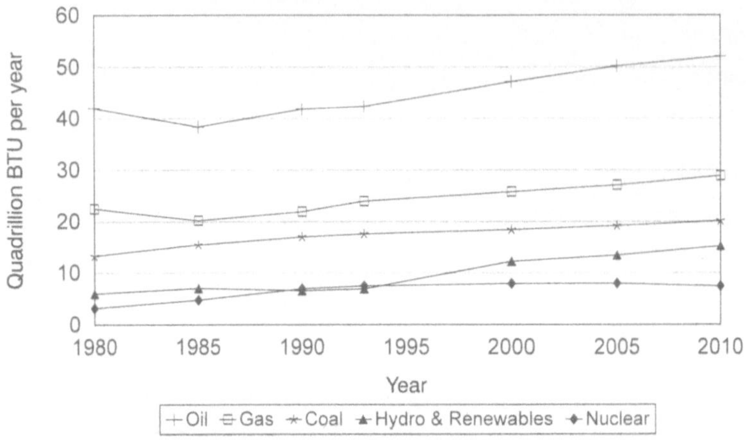

Figure 9. Energy consumption in North America.

The fuel use mix in Western Europe shows a decline in the use of coal due to environmental considerations (Figure 10). The most rapidly growing fuel is expected to be natural gas, growing at a 3.2 percent per annum rate between 1993 and 2010.

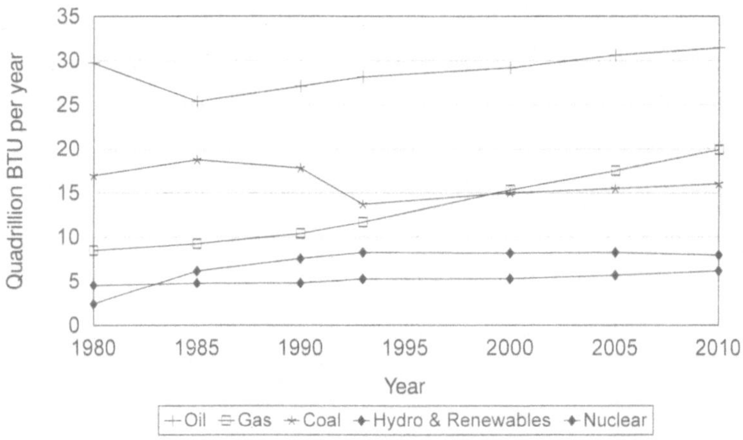

Figure 10. Energy consumption in Western Europe.

Similar to the occurrences in Western Europe, coal usage in Eastern Europe is expected to decline, and gas consumption is expected to rise (1.9% per annum between 1993 and 2010). Because it is indigenous to the region, natural gas is the most heavily used energy source (Figure 11). Russia produces nearly three-quarters of the natural gas in the region. Petroleum usage started to decline in 1990 due to economic considerations brought about by political changes but is expected to start rising again after 2000.

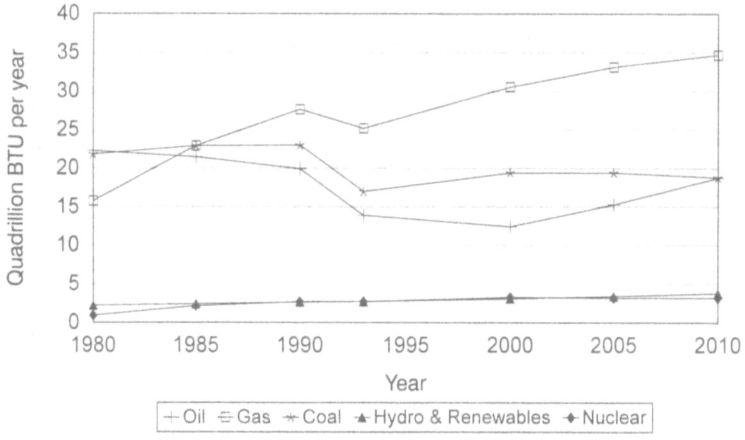

Figure 11. Energy consumption in Eastern Europe.

As mentioned earlier, the Far East and Oceania region is expected to have a rapid energy consumption growth of approximately 3.5 percent per annum. As shown in Figure 12, this energy is expected to be supplied by coal and oil almost exclusively. While coal consumption in the region is satisfied primarily through intra-regional production, well over half of the petroleum consumed has its origins outside the region.

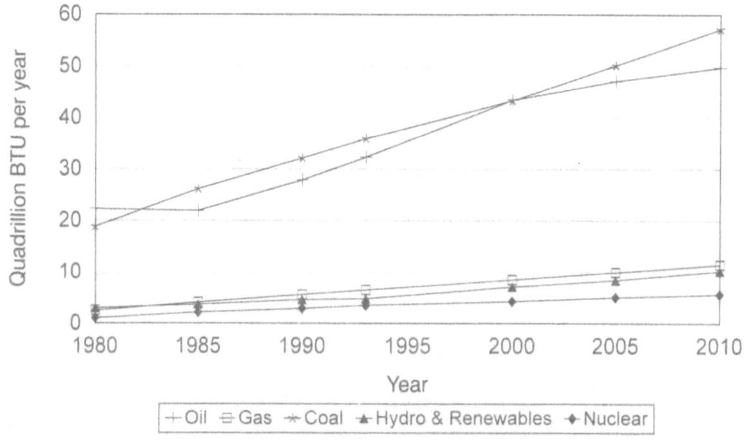

Figure 12. Energy consumption in the Far East and Oceania.

MISCELLANEA

With the continued reliance on fossil fuels, the question of resource depletion should be considered. As shown in Figure 13, current proven reserves of oil and gas will be depleted within the next 60 years at the 1992 production rate. However, if one uses the energy consumption growth rate projections for oil and gas shown in Figure 2, oil and gas reserves will be depleted in 32 and 39 years, respectively. As shown, there are substantial reserves of coal.

Given the growth in primary energy consumption shown earlier, it is instructive to consider the degree to which that energy is converted into electricity. Electricity generation, shown in Figure 14, and primary energy consumption, shown in Figure 3, have identical behavior. In both figures, the same four regions predominate, and the Far East region is projected to have tremendous growth. The growth in the Far East between 1993 and 2010 is equivalent to roughly 300 gigawatts of electrical capacity. This can be contrasted to the approximately 110 gigawatt growth in North America.

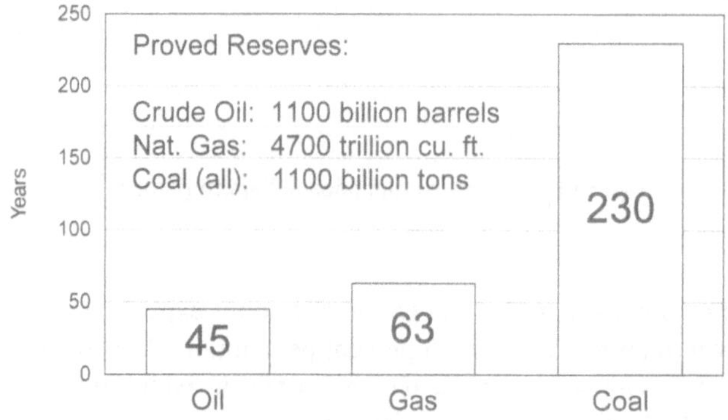

Figure 13. World fuel reserves when consumption is equivalent to 1992 production levels.

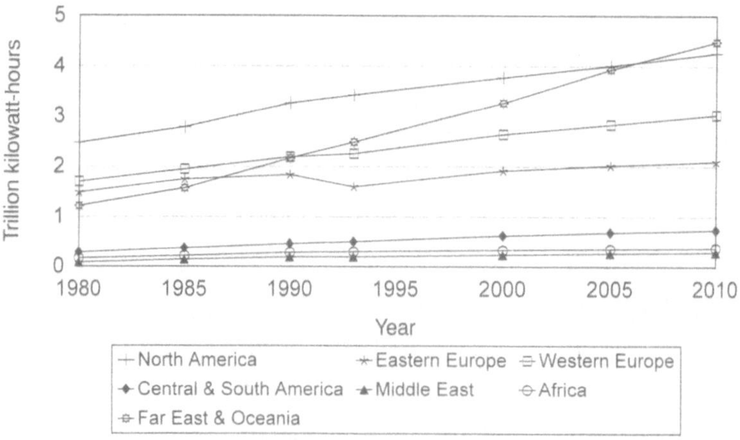

Figure 14. World electricity consumption.

With all the projected fossil fuel utilization, the adherence to the goals of the 1992 Conference on the Environment, also known as the Rio Earth Summit, should be evaluated. One outcome of the 1992 Summit was the desire to hold carbon emissions to their 1990 levels. Owing to the high growth in the consumption of both coal and oil, Figure 15 shows that the Far East is projected to have large growth in carbon emissions (3 percent per annum). Other regions are much closer to achieving the Earth Summit goal.

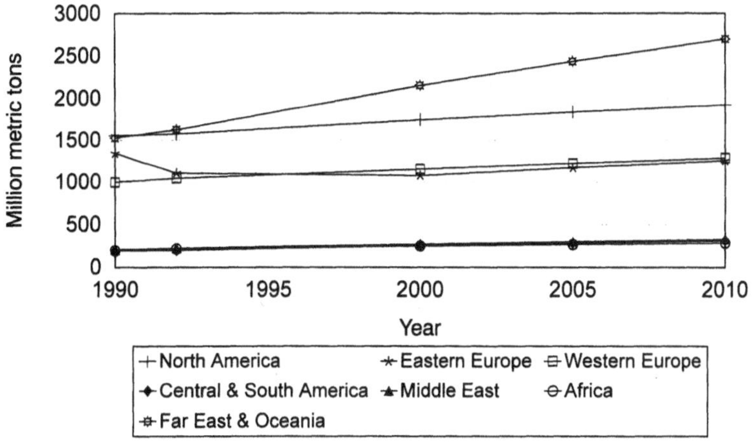

Figure 15. Global carbon emissions by region.

SUMMARY

This paper has sought to provide a global perspective on energy use. As shown in the various figures, oil has been and will continue to be the most heavily consumed fuel resource. However, current proven reserves may be depleted within the next 30 to 50 years. If that happens, the economics of supply and demand will produce a shift in the energy mix toward more plentiful resources such as coal, nuclear, and renewables. Natural gas may not be a long-term resource either as its current proven reserves may be exhausted within the next 40 to 60 years.

Of the seven major regions of the world, the Far East and Oceania region will experience the greatest growth in energy consumption, primarily through the use of oil and coal. Roughly 300 gigawatts of additional electrical generation capacity is projected for this region by 2010. Growing fossil-fuel combustion will increase carbon emissions in this region by 3 percent per annum.

ACKNOWLEDGMENT

The submitted manuscript has been authored by a contractor of the U.S. Government under contract No. DE-AC05-96OR22464. Accordingly, the U.S. Government retains a nonexclusive, royalty-free license to publish or reproduce the published form of this contribution, or allow others to do so, for U.S. Government purposes.

REFERENCES

1. *International Energy Annual 1993*, DOE/EIA-0219(93), U.S. Department of Energy, Energy Information Administration, May 1995.

2. *International Energy Outlook 1995*, DOE/EIA-0484(95), U.S. Department of Energy, Energy Information Administration, June 1995.

3. *World Energy Projection System Model Documentation*, DOE/EIA-M050(92), U.S. Department of Energy, Energy Information Administration, June 1992.

NUCLEAR PROSPECTS IN SOUTHEAST ASIA AND THE FAR EAST: A REVIEW OF EMERGING MARKETS FOR NUCLEAR POWER

Gerald Clark, Secretary General

The Uranium Institute
London SW1X 7LT

ABSTRACT

World energy demand driven by population growth and economic development will be over 50% higher in 2020 than in 1990. More and more of this energy is being provided as electricity. Nuclear power already provides a significant proportion of the electricity generated world-wide. World demand for electricity will double by 2020, but in the first 15 years of this period world-wide nuclear capacity will expand by less than 20%. Much of this expansion will occur in the emerging markets of the Far East and South East Asia, whose energy needs are expanding at a fast rate because of their position in the development cycle. An even larger role for nuclear power in these countries is possible despite financing difficulties. It is desirable, even inevitable, if they take the aims of the 1992 Rio Convention on Climate Change seriously.

The paper reviews present plans for nuclear power in this area, including those countries which at present do not have nuclear power but consider it a serious option for the future. It will consider the financial implications.

INTRODUCTION

According to the World Energy Council projections, world energy demand, driven by population growth and economic development, is expected to be over 50% higher in 2020 than in 1990, even when conservation and improved efficiency of use are taken into account. World demand for electricity is forecast to double by 2020.

The core business of the nuclear fuel industry is the generation of electricity. With the end of the nuclear arms race, there is only one large scale use for uranium: namely in nuclear fuel. Yet on present plans known to the Uranium Institute, world-wide nuclear generating capacity will expand by only 20% over the next 15 years. Compared to the total rise in the

Economics and Politics of Energy
Edited by Kursunoglu *et al.*, Plenum Press, New York, 1996

demand for electricity just mentioned this expansion of nuclear generating capacity looks modest, especially when, apart from hydro, nuclear power is the only source of electricity which has the ability to make a significant impact on controlling the growth of fossil fuel use and hence the build-up of CO_2, and other greenhouse gases in the earth's atmosphere.

Put another way, if present plans are not added to, nuclear's share of the world electricity generating market will decline in the next thirty years to about two thirds of its present level. If the world collectively wishes to moderate the potential for climate change attributed to CO_2 and methane emissions associated with fossil fuels, then it would do well to examine the possibility of nuclear's playing a larger role than is currently envisaged.

It is no coincidence that nuclear power is marking time in Europe and North America: electricity demand there has been growing only slowly for the past decade and looks likely to remain so. There is, for example, still a surplus of base load capacity in North America.

But in the developing world, and especially in the Far East where many national economies are expanding rapidly, the energy picture is quite different from Europe and North America. There energy saving, though full of potential in theory, takes second place to plans to increase supply. It is therefore not surprising that there is already significant growth in installed nuclear capacity in the Far East, especially in China, Japan, South Korea and Taiwan. The same is true of South Asia. It is equally unsurprising that the main present prospects for further growth in nuclear power are to be found in that great arc of territory which stretches round the coast of Asia from Japan to Iran.

Currently, there are 16 new reactors under construction in China, India, Japan, Pakistan and South Korea. Other countries such as Indonesia, Iran, Malaysia, Thailand, and Vietnam do not at present have nuclear power but have announced plans to develop it, as has Turkey, which is remote from the Far East, but has much in common with the countries just mentioned in energy development terms.

The aim of this paper is to review the declared nuclear energy plans of these countries, and to examine some of the constraints which affect their implementation.

GENERAL

There are a number of reasons why the countries mentioned have chosen nuclear power as part of the solution to their energy needs. Many of them are experiencing strong population growth together with rapid industrialisation and economic growth. A large increase in the supply of energy will therefore be required to sustain these growth rates. In most of these countries electricity has in large proportion been generated from fossil fuels: coal in the case of China, oil in the case of Indonesia, Iran, Malaysia and Thailand. However, with the exception of oil-rich Iran and Indonesia, oil imports adversely affect their trade balances. Even in Iran and Indonesia it is understood that the building of nuclear plants would eventually improve the trade balance by releasing oil for export, with significant effects on export revenue.

In addition, there is a growing desire in these countries to meet their electricity demand in an environmentally friendly way. It was notable that many developing countries spoke strongly at the Berlin Conference on Climate Change in March/April 1995, two years on from Rio, in favour of restricting the emissions from fossil fuels despite strenuous efforts from OPEC countries to prevent them. There is increasing evidence of willingness in many developing countries to diversify away from fossil fuels towards other more environmentally friendly forms of electricity generation, including nuclear power. In countries such as China and India, where transport and transmission distances are large, nuclear has advantages over fossil fuels. The World Energy Council has analysed in detail world energy/economic growth scenarios and reports that many countries consider a nuclear programme to be an essential part of their energy future.

I give in Annex A, a brief country by country analysis of short term (15 years) plans in the most significant countries in the Far East and South East Asia for the construction of nuclear power stations. In practice, apart from China, none of them has published anything more long ranging. There are three categories: those who long ago made the commitment to nuclear and already have a significant number of nuclear power stations (Japan, South Korea and Taiwan, in contrast with most OECD countries, have continuing plans to enlarge the nuclear sector); countries which have embarked on a nuclear programme, and have made some progress (China and Iran); and those countries which have only now reached the stage of economic development where it makes sense for them seriously to contemplate an investment in nuclear power. In both the last categories, there are clear signs that in each case serious consideration is being given by the authorities to exploiting nuclear as part of a balanced energy mix.

The first category apart, attentive readers of Annex A will realise that this catalogue of plans and firm intentions does not represent a quick fix in any of the countries concerned. The plans summarised there will do nothing for the immediate needs, and not so much for the medium term energy needs, of these countries. Assuming they can find ways of financing and thus of implementing these plans they will lay the foundations of possible expansion later. They will in effect reach the position which China has now attained 15 years after opening negotiations with western suppliers for the construction of a nuclear power station at Daya Bay. If nuclear power is to play a significant part in satisfying their medium and long term energy needs they will have to take some further positive decisions in the interim.

Even though the development of nuclear power in the emerging economies of the Far East is at an early stage a number of interesting patterns have already emerged. The CANDU technology is playing a notable part in more than one country, perhaps because of the flexibility of the design, and the fact that it does not involve the operator in the purchase of enrichment services. There is a growing Canadian presence in these markets. Korea and Pakistan use CANDUs, and Korea is building more. China is assessing them; so is Indonesia. Indian technology was originally developed in conjunction with AECL, although it has followed an independent track for many years. Away from the Far East, Turkey is looking at the CANDU for its first reactor.

Russia is playing a surprisingly active role. It has contracts to build in China and Iran. (China is also hoping to build reactors in Iran.) Both have exploited the US led Western boycott of Iran for political reasons. Russia is also building an enrichment plant, based on its version of the centrifuge technology, in China, and there has been talk of its building one in Iran. There are some signs that the old technical alliance between Russia and India could be revived. China is building a reactor in Pakistan and hopes to sell further technology and services.

South Korea is setting up deals using their basic design all over the place, not just in North Korea, but in Vietnam and especially in China. Although Western bidders were active in the latest Taiwan reactor contract they have yet to clinch the deal. Despite French successes in China, the mature Western PWR technologies are not having it all their own way in the one area of the world where the ordering of new power reactors still flourishes. The Korean basic design seems attractive to the Asian customer, and although based on Combustion Engineering technology may well supersede it as the Korean steel and shipbuilding industries have taken over from the models they originally imitated.

There are two sets of obstacles to the expansion of nuclear power in the emerging countries. One set is institutional or political. The other is financial. Some may argue that there is a third, namely technical, but if the first two can be overcome I suspect the last is not so difficult to deal with.

INSTITUTIONAL FRAMEWORK

Nervous politicians and commentators tend to assume that the spread of civil nuclear power increases the risk of the proliferation of nuclear weapons, regardless of the fact that all the existing nuclear weapons states developed their weapons quite separately from civil nuclear power, and in four cases out of five, before they did so. Moreover, since President Eisenhower's Atoms for Peace speech in 1953 there has been a generally accepted assumption that it ought to be possible to arrange matters so that civil nuclear power does not lead to the spread of weapons. The IAEA was established primarily to supervise such arrangements, and especially since the entry into force of the Non-Proliferation Treaty in 1970, an international system of supervision and inspection has grown up which is surprisingly effective given the difficulty of the task. From a slow beginning the system has become close to universal, and those countries which have chosen to remain outside it have found their nuclear progress more and more difficult as their ability to enjoy international co-operation has declined. The Treaty has become the basic system of international regulation and supervision of trade in nuclear materials and technology. It provides an assurance to all companies involved in it that their business will not be subjected to political risks which they cannot anticipate.

In May 1995 the Treaty was extended "indefinitely", establishing the international regulatory regime for civil nuclear trade permanently. This removes a major uncertainty affecting nuclear investment in developing countries. The strengthening of the international safeguards regime which was called for by the Extension Conference, and the subsequent moves by the IAEA to take measures to put it into effect, can only have the long-term result of making it easier, assuming that the members of the Treaty are seen to take their obligations seriously, to develop nuclear energy in countries which up to now have been unable to enjoy its benefits. These two strands should be mutually reinforcing. If the non-nuclear states see that strict observance of their non-proliferation obligations (under Article III of the Treaty) brings them the benefits of nuclear power (as a result of the application of Article IV) they will be the more inclined to apply safeguards rigorously. Such virtuous behaviour will make the transfer of technology and investment a more readily embraceable option by the supplier states.

FINANCIAL IMPLICATIONS

Availability of capital, both domestically and from international sources is in many cases the main limiting factor in the financing of nuclear power projects. In contrast with Western Europe and North America, or even the Soviet Union 30 years ago, developing countries usually lack the internally generated funds necessary to finance the construction of nuclear plants costing several billions of dollars each. Their funding for power projects must therefore be found externally. When seeking external funding one of the greatest threats to investment in new nuclear plant is competition for scarce funds from alternative investments, both within the energy sector and elsewhere. An added problem (which again did not apply in Europe 30 years ago) is the overall level of indebtedness of these countries, caused in part by previous power sector investments.

There are a number of possible solutions: utility type financing, joint ventures, and project financing of the build-operate-transfer type.

Utility type financing with government participation is one way of reducing the high cost of capital for new nuclear projects in emerging countries with no pre-existing nuclear power programme.

A joint venture involving the central government, local government and large private enterprises (domestic and foreign) is another possibility. The contribution required from each of these entities is usually in proportion to their respective requirements or demands on the completed project. Thus, central government investment would aim to ensure the power demands of its state-owned key enterprises, local government investment should meet the needs of the local economy and that of the private enterprises their own requirements. Joint ventures have been used successfully in China, and are under consideration in Turkey.

Private project financing based on the Build-Operate-Transfer (BOT) scheme may be difficult to achieve for first time projects due to the high return on equity (ROE) required to protect investors against perceived risks. But studies of the BOT scheme conducted in Indonesia by three consortia of nuclear vendors suggested feasible results and have been submitted to the Indonesian Government for consideration. Compared with conventional financing, the average electricity price to the consumer would be higher under this financing scheme in order to take into account the higher rate of investment return (ROI) required to protect against all risks.

Government commitment to support large infrastructure projects such as nuclear power plants by allocating a large portion of the available national resources is essential to establish sufficient confidence among external institutions and funding agencies such as the Asian Development Bank, etc. This will in turn encourage OECD governments and other lenders such as commercial banks and export credit agencies to support new projects.

Preserving foreign currency is seen by some countries as being of paramount importance and they are therefore keen to do business on a barter basis with Russia. They can supply food and clothing to Russia in return for Russian reactors.

In contrast, the South Koreans are able to pursue their ambitious nuclear power programme using their own resources. South Korea's latest nuclear power unit, Yonggwang-3, was its first nuclear reactor built without foreign assistance and went into commercial operation a few months ago.

China has, so to speak, now reached first base. Its economy has expanded enormously in the past 15 years, and the difficulties of capital formation as it was emerging from the financial desert of the Cultural Revolution have long since been replaced by the problems of success. Although the separate existence of Hong Kong as a hard currency area made Daya Bay a special case, nevertheless, the way it was financed may provide a model which other emerging countries could adapt to their own circumstances.

Guangdong Nuclear Power Joint Venture Company Limited (GNPJVC) built and operates a twin 900 MWe nuclear power station at Daya Bay, which commenced commercial operation in 1994. GNPJVC is owned 75% by Guangdong Nuclear Investment Company (GNIC), a subsidiary of China National Nuclear Corporation (CNNC), and 25% by Hong Kong Nuclear Investment Company (HKNIC), a wholly-owned subsidiary of China Light & Power Company Limited (CLP) in Hong Kong.

The Daya Bay project cost just under $4 billion. Ninety percent of the financing was arranged through export credits and commercial loans, with the Bank of China acting as guarantor to GNPJVC. French and British export credits were used to pay for plant equipment costing about $2.6 billion. During the 20-year period of the joint venture, electricity generated is distributed to the Chinese and Hong Kong partners in a 3:1 ratio, proportional to their equity holdings. GNIC re-sells 45% of its share of electricity to HKNIC for sales in Hong Kong. This earns foreign currency in order to pay off the loans.

After-tax profits are shared by the partners in line with the equity ratio after provisions have been made for operating costs, spent fuel and radwaste disposal, plant depreciation and decommissioning costs. At the end of the joint venture, the plant is turned over to GNIC who will continue to operate the plant for the remainder of its life.

CONCLUSIONS

This paper has reviewed the nuclear power plans of the emerging nations of Asia. These countries are on record as considering an expanded nuclear programme to be an essential part of their energy future, an energy future of surging electricity demand and a desire to conserve natural resources and to curb fossil fuel emissions. The financial implications of establishing such programmes is a key consideration, but not necessarily a total stumbling block.

Compared with the increase in energy demand in the countries under discussion their plans for installed nuclear capacity over the next 15 years look relatively modest. But compared with current prospects in most of Europe and North America the willingness of the emerging countries of Asia to include nuclear power in their energy policies is a daringly positive commitment. From small acorns do great oak trees grow. A similar review in 1965 of the nuclear plans of Europe and North America would have been similarly modest, yet 30 years later we have several countries with more than 50% of their electricity coming from nuclear and a much greater number where nuclear supply is greater than 20%.

The nuclear industry is now competing in a world which is very much aware that gas-fired combined cycle plants require far less in initial capital input, do not provoke opposition from the campaigning public, can be built in half the time it takes to construct a nuclear power station and, at present gas prices, can almost certainly deliver electricity to the consumer at a lower unit cost. Fortunately for it, it is also competing in a world which is increasingly aware of the disadvantages of burning fossil fuels. China for example in publicising its present energy plans emphasises the advantage which putting a sizeable portion of its generation expansion into nuclear power will bring in this regard.

As I pointed out earlier the plans adumbrated above (and described in greater detail in Annex A) only cover the next 15 years, a relatively short time scale in nuclear industry terms. If China is a good model, the really significant opportunities to meet the inevitable increases in demand will occur after that. But the decisions which will enable nuclear to play its just part will be taken during that time-frame. Nuclear will need to use every argument available to persuade governments and private investors to expand on the modest plans which this paper has surveyed.

REFERENCES

1. *The Global Uranium Market: Supply and Demand 1992-2010*, The Uranium Institute, 1994.
2. *Energy for tomorrow's world: the realities, the real options and the agenda for achievement*, 15th World Energy Council, Kogan Page, London, 1993.
3. 1992 Energy statistics yearbook, United Nations, New York, 1994.
4. *Energy statistics and balances of non-OECD countries 1991-1992*, OECD, Paris, 1994.
5. Ahimsa & Adiwardojo, The nuclear power programme in Indonesia, Proceedings of the 18th International Symposium, Uranium Institute, 1993.
6. B Gul Goktepe, Exploring the nuclear power option in Turkey, Proceedings of the 18th International Symposium, Uranium Institute, 1993.
7. Bhardwaj et al., Fuel design trends in Indian PHWRs, Proceedings of the 19th International Symposium, Uranium Institute, 1994.
8. A Langmo, A perspective of the evolving nuclear power programs in the Asia-Pacific region, Proceedings Volume 1, 9th Pacific Basin Nuclear Conference, Sydney, 1994.
9. Zeng Wen Xing, Guangdong Daya Bay nuclear power station project, Proceedings Volume 1, 9th Pacific Basin Nuclear Conference, Sydney, 1994.
10. Subki & Supadi, Recent progress in the feasibility study for the first nuclear power plant in Indonesia, Proceedings Volume 1, 9th Pacific Basin Nuclear Conference, Sydney, 1994.
11. Outlook on Asian nuclear power, *Nucleonics Week*, 30 June 1994.
12. *1992 International energy data*, World Energy Council, London, 1993.
13. H Murata, Peaceful uses of nuclear energy in Asia and role of Japan, Plutonium No. 8, Council for Nuclear Fuel Cycle, Tokyo, 1995.

ANNEX A

Country Analysis

China China's present installed nuclear capacity is 2.1 GWe. The stations at Daya Bay and at Qinshan have begun commercial generation. Contracts have now been signed with the Framatome/GEC-Alsthom consortium for the construction of two further 900 MWe units at Lingao, a mile or so from Daya Bay; construction to begin in 1996, and commercial operation in 2002-3. Four similar units are planned for erection on the Pearl River 60 miles to the west. The Qinshan design will be scaled up from 300 MWe to 600 MWe, and four such reactors are planned for the Qinshan site. Two are already under construction. There is an outline deal with the Russians for the construction of four 1000 MWe VVERs in Liaoning Province in the Northeast, contracts for two of which have recently been signed (construction to start 1996). The Chinese have also announced a plan for two 1000 MWe reactors in Shandong province, and for four in Fujian. They have also made a preliminary deal with the South Koreans who have the contract for the first scaled-up Qinshan reactor for the possible construction of up to 40 reactors along the China coast over the next 30 years. All these reactors are in rapidly expanding industrial areas far from the hydro resources of the Yangtse valley or the coal deposits of the North West. China expects to have 10 GWe installed or under construction by the year 2000, and 23 GWe by 2010. China is also building a centrifuge enrichment plant, using Russian technology; and has a new fuel fabrication plant, based on French technology, at Yibin.

Alone of the countries in this survey China has outlined its intentions with regard to nuclear beyond the next 15 years. Government spokesmen have talked of having 150 GWe installed capacity by the middle of the next century. Against present installed capacity (2 GWe) these figures look very ambitious, but against the likely demand for electricity from the booming Chinese economy they look modest, almost inadequate.

India India's electricity demand will nearly double by the end of the decade. Currently around 72% of electricity generation is from coal which involves burning annually 200 million tonnes of bituminous coal. This makes India the fifth largest producer of greenhouse gases. Currently there is a total of ten operating reactors, accounting for about 2% of India's electricity. A further six units are under construction for completion by 2001 and an additional eight units are planned for completion by 2010. But even assuming that construction of these goes ahead nuclear stations will supply only about 6% of India's total electricity consumption in 2010. However, a serious shortage of investment has delayed the country's nuclear programme to date, and it seems likely that this will remain a limiting factor. On the positive side India has its own uranium and large deposits of thorium. It does not require a SWU facility.

Indonesia Indonesian demand for electricity is growing by some 15% annually, creating an urgent need for additional electricity supplies. It is estimated that for the island of Java, which accounts for 80% of Indonesian electricity consumption, there will be a need for an additional 10 700 MWe of capacity by 2000. Under the government's energy diversification policy its aim is to reduce domestic oil consumption and to promote other energy sources. At the same time, the Ministry of Population and Environment has recommended that coal-fired capacity be limited to 15 GWe (equivalent to an annual 40 million tonnes coal burning limit for Java). Following recent studies which have shown that nuclear power is economically and technically feasible in Java and that the recommended unit size is about 600 MWe, the country's Atomic Energy Agency, BATAN, is engaged in further feasibility and site studies on the north coast of the Muria Peninsular. A decision on whether to proceed with construction will be taken in 1996 and, if positive, the Muria plant is scheduled to be operational by 2005. Longer term, a total of 12 units, to be brought into commercial operation between 2005 and 2015, are under consideration.

Iran Iran has long had ambitions to have a nuclear power programme despite its vast oil and gas resources. In 1974 the Atomic Energy Organisation of Iran was established to begin construction of 23 large nuclear power stations planned to begin operation by the mid-1990s. But by the time the Khomeini revolution took place in 1979 only two German-supplied 1300 MWe reactors were under construction, at Bushehr. Work was then halted. The reactors were later damaged during the war between Iran and Iraq. In 1991 the German government decided against resuming work on the unfinished reactors and told Siemens to withdraw from the project. Since that time the Iranian parliament has ratified agreements for nuclear co-operation with Russia and China: Iran has announced that it intends to complete the two units at Bushehr and to build two additional 440 MWe reactors on the same site with technical assistance from Russia, and plans to build two further 300 MWe capacity units again at the same site using Chinese technology. There has been some speculation that it has also opened negotiations with the Russians about the supply of an enrichment plant. The Russians have resisted prolonged United States Government pressure to desist from giving such assistance to Iran, and have now signed firm contracts for the completion of Bushehr. Assuming the plans go ahead, the first nuclear power plant in Iran will be completed by 1999. Resa Amrollahi declared in June that Iran aimed to provide 20% of its electricity from nuclear, which would mean its constructing 10 reactors over the course of the next 20 years.

Japan Japan is already the third largest civil nuclear power in the world, with 50 commercial reactors in operation in mid 1995, accounting for over 39 GWe installed capacity. Three more are under construction. Seven with a total installed capacity of 8 GWe will shortly start construction, two before the end of 1995, and five in 1996. Two more will start their planning procedures in 1997, with the aim of entering into operation in 2006. The Japanese have a target of 70 GWe by 2010, and hope to have 45.6 GWe installed by the year 2000.

Although Japan's original motivation to go nuclear was the search for energy security, the promotion of nuclear energy is also billed as the main instrument of Japan's policy on climate change. Japan aims to stabilise per capita carbon dioxide emissions at 1990 levels by the year 2000, and the planned increase in nuclear share outlined above is aimed at meeting future electricity requirements without further increases in per capita CO_2 emissions.

Korea The Republic of Korea too has made a major commitment to nuclear power. It already has 11 reactors in service, with five under construction, and two more planned, on which the contracts have recently been let. Korea is creative in its application of nuclear technology. It has purchased from Westinghouse, Combustion Engineering and AECL, and is now examining the synergy to be obtained from siting a CANDU next to a PWR, enabling the former which normally runs on natural uranium fuel to be fuelled directly with the spent fuel from the PWR. Due to the vigorous performance of the Korean economy it is now able to finance the construction of nuclear power stations without recourse to foreign financial assistance.

In recent years the Koreans have developed what they call the Korean Standard Nuclear-powered Reactor, which has enabled them to enter the market for supplying reactor technology to others. They already have outline deals with China and Vietnam, and are going to supply two units to North Korea as a result of the deal which has been struck between Washington and Pyongyang.

Malaysia Malaysia's demand for electricity will almost triple by the end of the decade, to about 13 GWe from about 5 GWe now. Against this background and a desire to diversify away from fossil fuel capacity, the government has announced its intention to develop nuclear power. Although it has an extensive programme of nuclear applications in various fields and is studying the nuclear power option to help meet projected electricity demands, the country's nuclear power programme is not as advanced as other countries' and no date has yet been announced for operation of the first reactor.

Pakistan Chronic power shortages are already occurring and the government is looking towards an expansion of its nuclear power programme to meet the growing shortfall. Nuclear accounts for 1% of the country's total electricity supply; hydro (52%), gas (29%) and oil (18%) account for the rest. Its existing reactor is a CANDU. Construction of a second reactor is underway at Chasma, Punjab province. Latest reports suggest that it may come on line in 1998, but as there have been delays in the delivery of equipment from China there must be some doubt over this date. Negotiations continue with the Chinese over Pakistan's plans for further reactors.

Taiwan Taiwan is a densely populated island endowed with limited natural resources. Electricity's share of final energy demand is forecast to increase from 22% in 1991 to 33% in 2010. Taking into account energy conservation, power management and co-generation efforts, the average growth rate of electricity demand is forecast at 6.1% between 1991 and 2001. This will require additional electricity generating capacity of 17 500 MWe to be installed during this period. An expansion of nuclear power, which already accounts for 32% of Taiwan's electricity, is viewed as essential because fossil fuels needed for electricity generation are imported. International bids for two new plants are being assessed, with completion scheduled optimistically for 2002 and 2004. This will bring the total number of reactors in commercial operation to eight, with further units planned for later in the decade.

Thailand Thailand's nuclear power ambitions date back to the 1970s when the International Atomic Energy Agency was requested to assist with its study of various plans for nuclear power generation. Interest waned in the late 1970s when large quantities of natural gas were found in the Bay of Thailand but has been revived by forecast growth in electricity demand of 8-10% per annum for the next 20 years. A tentative plan to construct 6 GWe of nuclear capacity with six units has been proposed by the country's Electricity Generating Authority, under which the first two units are scheduled to be commissioned in 2006. However, there have been reports that the Government is inclined to drop consideration of the nuclear option until 2010. Development of nuclear power is in any case likely to be slow, due to a lack of public acceptance and to pressure from the World Bank which has urged Thailand to consider cheaper alternative fuels.

Vietnam Growth in annual electricity demand in Vietnam has been lower than in the other emerging nations and therefore there is less pressure for additional electricity generating capacity. Current electricity generation is by hydro (76%) and coal. Meanwhile, nuclear activities are concentrated in non-power applications such as medicine, agriculture and food irradiation. Although the first research reactor in Vietnam started operating in 1963 and plans for constructing a nuclear power plant were first reported in 1985, preliminary studies by the country's Atomic Energy Commission and the Ministry of Energy concluded that nuclear capacity of 800 -399 1000 MWe would be needed by 2010 at the earliest. In mid 1995 some acceleration of these plans was reported. Korea has offered to assist the Vietnamese with the construction of a 1000 MWe reactor, with construction starting in 2000. No final decision is expected before 1998.

THE ROLE OF SUBSIDIES AND PRIVATE INVESTMENTS IN SUSTAINABLE RURAL ELECTRIFICATION

Phillip W. Covell and Richard D. Hansen

Enersol Associates, One Summer Street
Somerville, MA 02143, and

José G. Martín

Department of Chemical and Nuclear Engineering
University of Massachusetts Lowell
One University Avenue
Lowell, MA 01854

ABSTRACT

From several points of view (environmental, resource conservation, financial, and social) the widespread use of photovoltaic (PV) systems for rural electrification is desirable. In this presentation, it is argued that full cost recovery is a key factor to ensure widespread PV use, and that temporary equipment and financial subsidies can actually hurt the rural market penetration of PV technology. If, as argued, full cost recovery is a key factor to ensure widespread PV use, developers and planners should adopt pricing strategies that reflect market conditions, and development assistance should focus on risk reduction through planning, technical assistance, training, and policy making.

1. INTRODUCTION

Decisions which are been made now on energy options will affect all us for decades to come, and those decisions are based on present perceptions of benefits and risks. There are limitations to our cognitive abilities when trying to evaluate complex systems which vary in time, however, and global energy consumption and supply are examples of such a system.

A major change is underway, and energy demand is increasingly being determined by developing countries of Asia and Latin America. The reasons for this are several: population growth, industrial development, and urbanization are important factors. The consequences of the ongoing shift will be enormous: they will place heavy demand on financial resources, accelerate fossil fuel depletion, and will affect the environment. In particular, developing country plans rely heavily on the burning of low-quality coal, and this reliance will increase greenhouse gas emissions alarmingly.

Sustainable energy sources, and in particular solar and nuclear power, could help meet the demand without increasing those emissions, but several obstacles, including high capital costs and the *de facto* stigmatization of these technologies constitute formidable obstacles. Technologies are stigmatized when they are rightly or wrongly perceived by the public to be too dangerous. An outstanding example is nuclear power. There is another type of stigma that arises when a technology is somehow not taken as a serious option by investors and policy makers: "are renewables anything but hot air?" It may be argued that the (sometimes unreasonable) perception that solar power is not financially sustainable slows down the market penetration of the technology. Of course, the issues of stigma and costs are not independent: for different reasons, stigma makes both nuclear and solar options more expensive.

1.1. Cognitive Limitations

Psychologists are beginning to pay attention to the unsuitability of the human mind to develop appropriate images of complex systems. Whether we are forging images of ecological systems or of a global system of energy generation and consumption, these systems are complex and not transparent. Furthermore, their time evolution goes through quasi-stationary phases where our cognitive capacities falter.

When facing such systems, our minds tend to commit characteristic errors. For example, although the behavior of a system may depend in a complex way on an underlying network of interrelated variables, we tend to propose a reductionist hypothesis, where everything depends on a central variable.

Also, if a system evolves relatively slowly, we fail to develop a correct image of the system behavior. (Dörner (Ref. 1)) Instead of interpreting a time-dependent observation as an evolving one, we tend to look at it as a "state". Carbon dioxide concentration in the atmosphere has increased from 270 ppm to about 350 ppm in the last 150 years (Ref. 2) - but we take the present concentration to be "normal." Energy consumption in China has increased by a factor of 22 since 1952 (Ref. 3) - but we do not interpret this change as important. For us, these are "slow changes." We create the illusion that the present is the normal state of the atmosphere, and that China's contribution to the greenhouse gases is not important. Because of the limitation of our minds, we tend to implicitly assume that the present is the way it always was, and always will be. Of course, we may be very surprised when the time evolution suddenly accelerates.

Change is accelerating in energy generation and consumption. The superpowers which expensively fabricated plutonium and highly enriched uranium during the Cold War, are now trying to find ways to get rid of them (Ref. 4.) Past predictions of natural gas shortages have certainly proved pessimistic (Ref. 5.) Predictions of crude oil price increases have been wrong indeed - in constant dollars, crude oil prices have nearly returned to their pre-1973 levels (Ref. 6).

Between now and the year 2020, according to some estimates, as much new power generating capacity will be installed around the world as was built in the entire past century. (Ref. 7) Where that generating capacity will be built constitutes one particular dramatic change. A majority of that new capacity will serve the surging economies of Asia, Latin America, and Africa. This change has tremendous social, environmental, and financial implications.

Recognition of our own limitations suggests that we stay away from reductionist solutions to the challenges posed by changing global energy demands. It is possible that natural gas reserves are immensely larger than earlier predictions, but it is not wise to assume that combined-cycle gas powered cogeneration plants will carry us smoothly until the middle of next century, at which time sustainable energy sources will somehow have overcome their political and financial obstacles. We argue here that some of those obstacles can be overcome now. For example, well-meaning subsidy policies may be making it harder for investors to commit to solar rural electrification - and this obstacle is one that can be overcome.

1.2. The ongoing shift

In the early 70's, the world was shocked to realize that control of oil had fallen outside of the industrialized countries. It took a decade, a second oil shock, and a world recession for the so-called developed countries to somehow adjust to that shift. Whenever possible, other fuels were substituted for oil, and conservation became a priority.

In the 90's, another major change is occurring: energy demand is increasingly being determined by developing countries of Asia and Latin America. Although the full consequences of this new shift will not be felt for many years, they are potentially greater than those arising from the crisis of the 70's.

According to market studies and forecasts publicized by General Electric, (Garrity, Ref. 8) generation in Asia is increasing at a rate higher than 5% per year, and during the next ten years, the Asian market will be twice the size of the North American market, constituting about 40% of the new orders for power plants. During that time frame, GE estimates the world-wide "uncommitted need" to grow by 3% per year to a total of 630 GW (this is close to the present U.S. generation, which is 700 GW) Of this total, 238 GW will be in Asia (growing at 5 - 1/2% per year).

According to "high growth" scenarios postulated by the World Energy Council, annual world energy demand could be double by the year 2020. According to the Council, this growth is dominated by Asia and Latin America, where demand will triple over the next 30 years. By comparison, demand in North America will rise by only 13%, so that energy use in the developing countries will account for as much as 60% of the world total, compared with 30% in the OECD (Ref. 2). China and India, home to 40% of the world's population, account for a large share of this predicted demand. Consumption in China and India have already grown prodigiously, albeit from a low level. In China, consumption has increased by a factor of 22 since 1952. India's power generation capacity has increased from 1,712 Mw in 1950 to 42,000 Mw in 1986 - this represents a rate of growth of 9% per year, which exceeds the growth rate of the GNP (Ref. 9)

According to EPRI (Ref. 7) China and India are expected to account for one-quarter of the projected growth in electric generating capacity over the next 25 years: each of these countries expect to build as much or more capacity in the next decade as was commissioned in the United States during the 60's - the last era of American grid expansion. China's government has announced plans to double the present installed capacity by the year 2000 - this corresponds to adding the equivalent of a medium-sized fossil fuel power plant every week or so. India's government estimates that the country needs to triple the present installed capacity in the next 15 years.

Demand and capacity will continue to grow everywhere - but in particular they will grow in the developing countries. The two awakening giants, India and China, are expected to account for one-quarter of the projected growth in electric generation capacity over the next 25 years (Ref. 7)

How good are these forecasts?

If anything, the scenarios look to cautious, not too bold. The Economist article quoted above tells about one planner working in Hong Kong for a big oil company "who is frustrated that colleagues back at head office routinely tell him that his Chinese forecasts are too big to be believable." For the sake of credibility, the forecasts are negotiated down to safer levels. "What emerges certainly looks more plausible, but it may also be wrong." The same article quotes an employee working on scenarios for the International Agency: "the numbers just went crazy...they were inconceivable in the real world." According to The Economist, "when it comes to scenarios, what you see is what their authors dare to think." (Ref. 3)

Regardless of the quantitative accuracy of the predictions, there is a major shift underway: energy demand in developing countries is being driven by powerful factors. Two of these are population growth and industrialization, which have been lucidly addressed in this conference and earlier ones (Refs. 10, 11, 12, 13). Urbanization is a third, closely related factor.

1.3 Urbanization

With industrialization, farm workers leave the land for the cities. In 1950, 29% of us lived in cities; today, 45% do - this is more than two billion people. Every year, twenty million people migrate to the cities - mostly in the developing countries. There are twenty "megacities" with more than eight million people - of these, fourteen are in developing countries - this represents more than two billion people.

Massive urbanization is a world-wide phenomenon. Brazil is an impressive example. In this huge country, where there are millions of square kilometers of barely inhabited land, 85% of the population lives in cities. If you fly anywhere in Brazil from abroad, you are likely to land in São Paulo, where you have a view of tall buildings stretching all the way to the horizon - Before this century is over - i.e., in five more years - it is expected that São Paulo will have more than 23 million people.

Will that make São Paulo the largest city in the world by then? No! By the year 2000, there will be 25 million "chilangos" (this is the endearing name that Mexicans give to those who live in *their* capital.)

There are many factors driving mass migration - and it has been argued that the economics of electricity generation and demand are not a negligible one. In the Dominican Republic, tellingly, internal migrants are said be escaping the *"campo negro y oscuro"* - that is, they leave the black and dark countryside for the (electric) lights of the cities. It just so happens that electricity supply in Dominican cities is not particularly reliable, but unreliable electricity is better than no supply at all. Most developing countries subsidize electricity generation - and the net result is that all the population - poor country folk included - wind up subsidizing the urban electricity consumers. Thus, at least in relative terms, they may wind getting poorer. They may not be aware of how their status is being affected by energy policy, but they do somehow get the impression that they -and their children! - will get a better survival chance in a crime-infested dirty city *"favela."* It is reasonable to argue (Refs. 14, 15, 16) that alternative decentralized rural electrification schemes may help slow down this massive migration.

Electrification and urbanization may be considered as constituting a system with positive feedback, because urbanization demands more energy consumption. At any income level, city people consume more energy than rural populations. More energy is required to take people from and to work, make and distribute products, etc. A study by Jones (Ref. 14) on the effect of urbanization on energy use in developing countries concludes that if India and China were to double the proportion of their population living in cities, energy demand would be 43% higher than now even if their national incomes and populations remained the same.

In conclusion, population growth, industrialization, and urbanization are major factors driving the ongoing major shift towards a world in which the developing countries determine the shape and size of world energy demand. What are the consequences of this shift?

1.4. Consequences of the Global Trends

1.4a Environmental Consequences

The means by which the energy demand of developing countries is met may have grave consequences for the global environment. As noted above, India and China, with less than 10% of today's demand, plan to build what would amount to a quarter of the world's new capacity by the year 2020. We may speculate on the impact on the environment of the energy plans for these countries.

Let us consider India. Whatever solution India adopts to meet its energy needs will affect every human being on earth. India is still in the early stages of development. The energy requirements to maintain development are huge, and meeting those requirements a major challenge. The country depends mostly on imported oil stocks for power generation, but several "orthodox" analyses from policy makers, sociologists, and economists who have considered different possibilities have concluded that burning coal is the only solution which is economically attractive for India (Ref. 17.) This seems logical, because coal is cheap and abundant in India, and it has been argued that limiting CO_2 emissions in a developing country like India inflicts too heavy a penalty on the future living standards of the rural population. The authorities may decide to support (i.e., subsidize) centralized models for power generation based on coal-fired power plants with long transmission lines.

There is a problem with this solution. If future development is based on coal power, extrapolating to a future economy where a billion people have of standards of living to which they legitimately aspire leads to a qualitatively different world environment. As environmentalists and other concerned groups have repeatedly pointed out, burning of coal represents a grand scale version of the "Tragedy of the Commons" - and a billion "shepherds" own a lot of sheep. Coal burning is a major contributor to the building of carbon dioxide (CO_2) levels in the atmosphere, whether a coal process is classified as lean burning or not. Other gases and solid emissions will result from a coal-based economy. The long -term disadvantages of a development model emphasizing the mining and burning of coal will easily offset economic and any other desired benefits of any short term program.

The air in major Indian cities such as Calcutta and New Delhi is already more polluted than in cities of western industrialized countries. According to the World Health Organization guidelines, the concentration of suspended particulate matter in air should not exceed 230 micrograms per cubic meter more than seven days in a year: in the above referred Indian cities, those levels are exceeded more than 200 days a year (Ref. 18)

(The problem of air pollution is worse in other developing countries. A grim joke in Mexico City is that one should not trust air one cannot see. In Bangkok, according to a study by the United Nations Environment Programme, traffic jams cost $1 billion a year in medical and other side-effect. On cold winter days, the smog blanketing China's coal-powered northern cities is between six and 20 times the highest levels in the West.)

Much of the coal available in countries such as India, China, and Brazil have high ash and/or sulfur content, and mitigating the environmental impact of the use of such coal is expensive. Although the immediate local environmental consequences of using that coal are grave, it is conceivable that a richer India or a richer China may afford the technical fixes to mitigate the local problems: after a certain income threshold, people become environmentally aware, and somewhat willing to pay for environmental fixes. There is no fix for the global impact of gases such as carbon dioxide, the main contributor to global warming.

According to a study of possible scenarios released last year by International Energy Agency, the Indian and Chinese emissions of carbon dioxide by the year 2010, will account by as much of a quarter of global carbon emission - and will be almost as big of those of the *whole world* in 1980. Chemistry dictates that the amount of carbon dioxide emitted in coal burning is close to twice as much as that emitted in natural gas.

By 2025, according to a pessimistic scenario reported by the Journal of the Electric Power Research Institute (Ref. 7) 70% of worldwide carbon emissions will originate in the so-called developing countries. Much controversy surrounds the issue of the greenhouse gases, and their effect on global warming. In the next two decades human beings will perform a major experiment to resolve that controversy.

There are energy alternatives which do not release carbon dioxide. Hydroelectric power is one - although the methane generated by plants alternatively thriving and drowning behind dams in exuberant climates in South America constitute a major contribution to greenhouse gases. Nuclear and solar power, and conservation, are other alternatives.

1.4b Fossil fuel depletion

Oil is cheaper now that at any time since 1973, and there is a natural tendency for many observers to feel confident about the supply and price of oil. Although predictions are risky, it is fair to wonder whether those observers may be in for an unpleasant disappointment.

There may be enormous untapped oil reserves in the Tarim Basin in the Takilamakan Desert in China, in former Soviet republics, or under the seabed off the coasts of South America and Vietnam, but formidable challenges attend tapping those reserves and others, and one the most formidable challenges is raising the needed money. More ominously, there is a huge uncertainty in demand. What will happen to demand if western economies prosper, or if Russia recovers? More ominously, what will happen to the demand from the developing world. The World Energy Council predicts that Asia and Latin America will consume 25 m/b more in 2020 than in 1990, but this is only a guess. Richer Indians or Chinese will consider buying an automobile. How many will be too many? Until 1993 China exported oil - now it is a net importer, and fuel for its new cars will soon be arriving in super tankers from the Middle East. How much oil will China eventually import?

1.4c Financial consequences

The financial resources needed to create the electrical generating capacity to meet demand in developing countries are enormous. The World Bank has predicted that more than $1 trillion will be invested in those countries in the 1990's. The investment needed for China to sustain its planned rate of electricity capacity growth will cost roughly 3% of its gross domestic product each year - and China's electric utilities do not have the money. Neither do the state electricity boards in India, which are technically insolvent.

Lenin defined communism as "Soviet power plus electricity". It is somewhat ironic that electrification in the developing world is turning into a stimulus for global capitalism, because developing countries are turning to private capital to build capacity. In fact, Michael Jordan, head of Westinghouse, has been quoted as referring to the response of private investors as a "feeding frenzy."

Private investment may help meet the huge capital demands required - and help introduce some market discipline to counteract a natural political tendency towards subsidies and inefficiencies, but the road ahead is a challenging one. The Enron impasse in India reminds us how hard is to change established policies, and Indian stock markets seem to rise and fall according to the world's perception of the resolution of that impasse. Another problem is that investors (and governments) are haunted by memories of the debt crisis of the 1980's. About 40% of the money lent to developing country governments then was to be invested in energy projects, and some environmentalists claim that loans to developing nations have led to a spiral of debt and environmental degradation. (These claims have been refuted by analysts who argue that such a connection cannot be proven (Ref. 19.)

We shall be reexamining how the financial obstacles to solar rural electrification and the role of government subsidies. First, however, it may be of interest to review how different perspectives affect our views of possible solutions to the energy challenges.

2. VIEWS AND PERSPECTIVES

One of the effects of the "first energy shock", in the 1970's, was a certain obsession with energy conservation. Although the industrialized countries eventually adjusted to that shock, its effects, positive or not, are still felt.

The U. S. electrical generation industry is certainly not what it used to be. Mr. James D. Shiffer, Executive Vice President of Pacific Gas and Electric Company, has presented a picture of the present and future of the U.S. Electrical Generation Industry". He is pessimistic about the role of nuclear power in the U.S. was noted above. His company, PGE, is meeting the demand through a combination of demand side management, cogeneration, independent power generators, and alternative energy sources. His company supports retail wheeling, and he personally looks forward to the day when "a little old lady in Pasadena could purchase power from an independent producer in Coral Gables, if that producer was the least-cost producer". He referred to the problems associated with the investment in transmission lines, and saw the conversion of the utility industry to one of "franchises" to transport electricity.

Over the ensuing years, a veritable schism have developed amongst different analysts, and nothing escapes this schism: the desirability of growth, the fears of innovation, the interpretations of individual freedom, and the roles of the markets.

On the one hand, there are strong proponents of sustainability. (See, for example, Rosenfeld (Ref. 21)) According to some proponents, "sustainable energy" is one where irreplaceable fuels are not depleted and the environment does not continue to degrade. The switch from "supply" to "demand" side management makes it possible for the U.S. economy to make a transition from one which is based on traditional polluting fuels to one which is powered by "sustainable energy".

This vision is not shared by all observers in the developed countries. For example, Dr. Edward Teller (Ref. 20) has publicly lamented the proliferation of literature based on pessimism related to the effects of productive activities, and the sloppiness of our social and even professional consideration of effects such as radiation. Dr. Teller asserts that energy policy is being over influenced by fear, such as fear of radiation at low levels, and this fear is used as an excuse for planning - and ""'overplanning"".

A similar schism seems to be developing among analysts of developing country needs. Steven Wiel (Ref. 22), of the Washington Project Office, of the Lawrence Berkeley Laboratory makes emphasis on sustainability by using as example the need for lighting in the developing country such as Nepal, and compared the alternative of building, fueling, and maintaining the generating capacity needed to provide lighting through traditional incandescent lamps, with the alternative of using fluorescent lights. According to Mr. Wiel, it is clear that the latter choice is the right one, and financing institutions and governments should take the measures which may be necessary so that the wise choice is made.

According to Mr. Wiel, a traditional approach tends to impoverish the developing country, wasting its financial resources in the subsidy of unneeded capacity, the purchase of foreign fuel, and the destruction of resources. For Mr. Wiel's, "sustainable energy" is one for which the economy can generate the resources to sustain itself - and this generation is made possible through a change from an emphasis on energy generation to one which emphasizes the provision of energy services. "Sustainability" can be defined by capital constraints, without environmental arguments.

The Electrical Engineering Research Institute appears to agree that renewable energy resources are considered directly cost-competitive with conventional alternatives in far more applications in a country such as India than in countries with highly developed electricity supply infrastructures (Moore, Ref. 7). Other authors have presented similar arguments (Ref. 23).

Because developing country needs are so huge, any renewable energy contribution will help: the more the better. It does seem that such contributions will pale in contrast with the huge build-up of traditional (i.e., fossil-powered) electric capacity unless the means are found to overcome institutional and financial obstacles. To overcome those obstacles, it may be helpful to examine what underlying factors influence what passes for "rational" decision-making in energy policy.

2.1. Technological stigma

In ancient Greece, a stigma was a mark placed on an individual to signify infamy or disgrace. Although technology has the perceived potential for good, it also poses risks. This second face of technology has brought about a generalization of stigma: technologies or products are being stigmatized when they are rightly or wrongly perceived by the public to be too dangerous. Outstanding examples are those technologies involving the use of chemicals and radiation.

Stigmatization de facto eliminates technological options, and so it is important to understand the phenomenon. Gregory et al. (Ref. 24), who have analyzed the phenomenon, note that stigma often arises from some critical event, accident or report of a hazardous condition which sends a strong signal of abnormal risk, linking negative imagery and emotional reactions to the technology and motivating avoidance behavior.

According to Gregory et al., there are several features which stigmatized technologies share. For example, the source for the stigma is a hazard with characteristics such as dread consequences and involuntary exposure, that contribute to high perceptions of risk. The imagined impact are unbounded, in the sense that their magnitude or persistence over time is not well known. Public evaluations of advanced technologies tend to be ambiguous and inaccurate, and thus they contribute to their stigmatization. Stigmatization often results from the overturning of what had been perceived to be a positive condition: this may happen, for example, when a certain technology has promised cheap, safe power, and fails to deliver. It is perhaps not surprising that nuclear energy is being stigmatized, reflecting public perceptions of high risk, disappointment of failed promises, and distrust of authorities.

Another type of stigma affects options which are somehow considered irrelevant rather than dangerous. Policy makers and even investors may give lip service to such options, particularly if they are environmentally benign, while implicitly discounting them as being impractical or too expensive. In fact, they are discounted as so much "hot air". Such options carry the stigma of being "Quixotic" - although few take the trouble of reviewing the economic assumptions and the political and infrastructure framework that make them appear to be too expensive. There is no public outcry against such options, and in fact policy makers may be willing to subsidize such options on a token basis, reinforcing the image of an impractical alternative that requires artificial propping. Such an option may be an alternative for the future - and if we do not recognize and address the stigma issue, it will always be. It may be argued that solar energy collection and conversion systems are being stigmatized in this way.

3. BREAKING THE GRIDLOCK

Because of the limitations of the human mind, important decisions may be often made on the basis of reductionist views. Such decisions may be costly. In Nicomachean Ethics, Aristotle described "virtue" as "a state concerned with choice, lying in a mean, relative to us and determined by a rational principle, the principle by which the man of practical wisdom would determine it." We may consider applying this definition to a virtuous analysis of energy alternatives.

By necessity, any conclusion based on this definition must be of a general nature: Aristotle himself warned us that "in the sphere of action as in medicine or navigation, any general account must lack detail and precision." However, the general principle should be to avoid reductionist views of the alternatives. As a corollary, we should realize that different problems have different solutions. A good investment portfolio is a diversified one, and ecology and human social and economic systems are made robust by diversity. Conservation, cogeneration, renewables, and nuclear energy, can - and should - make a contribution to improve man's lot.

Nuclear energy and improved energy efficiency for fossil-powered plants have been discussed at quite some length elsewhere in this Conference. This paper concentrates in solar energy applications and photovoltaics (PV), and in particular in the use of photovoltaic system for sustainable rural development in the developing world. There are social and environmental reasons why widespread penetration of photovoltaic technology in the rural electrification market in developing countries is desirable as part of a diversified, non-reductionist approach to help meet rising global energy demands. With this background, let us discuss how can market interventions be structured to ensure long term benefits to mitigate risks for financial institutions, investors, and firms in the photovoltaic industry.

3.1 Full Cost Recovery

We have argued above that widespread use of PV for rural electrification is a desirable goal. A major shift is underway: energy demand is increasingly being determined by developing countries, with very important environmental and resource-conservation consequences, particularly when many of the exploding giant economies are planning their development on the basis of the burning of fossil fuels and in particular low-quality coal. A non-reductionist diversified approach is certainly preferable - one that includes solar rural electrification.

Our main interest here is in the financial consequences of the ongoing shift. We have seen the opinion of some analysts on the resources required for power generation to meet the exploding demand in the developing countries. According to Lysen (Ref. 25), *"Governments and utilities in developing countries have major plans for expansion of their power sector, which is estimated to require not less than a trillion dollars in the 1990s to achieve an increase of 384 GW in total power supply capacity (from 471 GW in 1989 to 855 GW in 1999). About 40% is in foreign exchange, and it is clear that the required US$40 billion /yr cannot be mobilized, even by the large multilateral banks (the present level of World Bank lending for the power sector is around UDS$7 B/yr."*

This will require private investment . According to Smith (Ref. 26), Frank, ("Policy Reform and Sustainable Energy Development, "Consultation on Renewable Energy and Sustainable Development in Latin America and the Caribbean (1994))

> *..."Getting from here to more efficient, more equitable and more environmentally sound practices is the project and policy challenge of the coming decade.*
>
> *...This challenge needs to be attacked at a time when the developing world needs an estimated $US one trillion over the next decade to keep its electric supplies at pace*

with economic growth. This nvestment goes well beyond the scope of the primary lenders and will require the growth of private sector financed power."

As the financial institutions look at the electrification challenge, they realize that for a technology to be financially as well as technically sound, its implementation requires full cost recovery. Thus, the technology requires a high quality of service, in the sense that it cannot tolerate a high number of failures. This in turn requires not only that the systems be robust, but also that the social, labor, and financial conditions exist for the technology to operate reliably so that the systems are maintained and repaired when they do fail.

Electrification will require private investment - and private investment depends on full-cost recovery (FCR). According to Silverman and Worthman (Ref. 27), *"Investment and innovation in renewable energy - a key element of any long-term energy strategy - will not occur unless developers and investors perceive the industry to be attractive over time. The long term attractiveness of renewable energy industries should be maintained, for the sake of jobs, energy security, the environment, and our international competitiveness." Since the 1980's private investment (in renewables) has languished, scared off by early project failures, overoptimistic cost estimates and diminished government support."*

4. PRIVATE INVESTMENTS, NATIONAL POLICY, AND SUBSIDIES

When discussing the evolution of the human brain, the influential Hungarian-British writer Arthur Koestler once pondered rhetorically why Mother Nature had not invented the equivalent of a wheel? Koestler himself thought that the answer lay in the fact that there is no *continuous viable path* for an organism to create the wheel. Let a paramecium or a Tasmanian devil evolve in such a way as to start growing a wheel, and the creature will become extinct before the wheel becomes functional. Examples can be found among human endeavors (transportation, city planning, etc.) where utopias can be imagined but not reached through continuous paths. Conversely, the existence of continuous viable paths help explain the otherwise curious configurations reached by creatures and human creations.

Will solar PV ever make a really important contribution to help meet global energy demands? We may argue that one of the main reasons why it will is that there is a continuous viable path from where we are to the point where that contribution is important. There are already markets where the photovoltaic option makes economic sense. As the need in those markets is met, PV production increases, with implications in lowered mass manufacturing costs and in a stronger infrastructure. The lowered costs and better infrastructure open new markets; then, production increases further, and so on.

That PV technology makes sense for rural electrification throughout much of the world is fast becoming a truism to policy makers and development planners, particularly at high - and financial conservative - levels. The technology has environmental advantages, and costs advantages, and its implementation in rural areas seems easier than that of conventional technologies. PV is considered seriously by those comparing options. Financial and development institutions are evaluating how to advance PV market penetration in the vast areas of the world which are not served by central grids.

The problem is not whether PV will make an important contribution, but how can development policy facilitate market penetration of the technology. In other words, how can market interventions be structured to ensure long-term benefits to mitigate risks for financial institutions, investors, and firms in the industry?

This introduces the concept of subsidies, which can be interpreted in a fairly subjective manner as something governments "give" in order to promote any desired policy. According to Lazonick (Ref. 28) *"The prime objective of national economic policy is to contribute to continuous improvements in the material standards of living of the population.*

To achieve this objective, national economic policy must promote sustained economic development....

...the innovation process is, by definition, a learning process... Learning requires developmental investments in the productive capabilities of both the people doing the learning and the physical resources ... with which these people work. Developmental investments must be strategic.

The willingness of business people... to invest in the productive resources that can generate innovation will depend on how they evaluate their ability to overcome productive and competitive uncertainty."

It can be argued that there is always an element of subsidy in any innovative process. Rural electrification in the U.S.A., and nuclear power development everywhere are clear examples of subsidized innovative processes in energy. Even the young entrepreneur who works long hours at below-market rates to start a renewable energy business is subsidizing that business.

The argument may be carried further: there is always an element of subsidy associated with any energy policy. Coal is subsidized in Europe and Asia; hydroelectric power development is subsidized in Brazil, kerosene in India. It does not take much effort to identify an element of subsidy in any energy path - even in cases where government policy is ostensibly noninterventionist, there is a (frequently unrecognized) element of subsidy through external environmental and social costs which may not show up in traditional accounting practices.

If there is always an element of subsidy, it makes sense to structure market interventions so as to further desirable goals. There are abundant arguments for photovoltaic subsidization of solar-based rural electrification in comparison to grid extension or fossil-fuel based options. Those arguments relate to:

1) environmental quality
2) social development
3) economic improvement, including long-term foreign exchange implications
4) social development
5) political expedience, etc.

If a country is committed to rural electrification, as was the U.S. in the 30's and 40's, decentralized PV should be considered as a sensible and economic alternative to conventional rural electrification. It is often the last cost option even on a first-cost basis.

Mexico provides subsidies of 80% or more to PV systems for the rural poor: this commitment appears to be a permanent one as a matter of policy. The social and environmental benefits are obvious. The policy may be costly, and the market penetration be limited by the cost, but it is clearly hard to argue against such subsidies, which appear to constitute a permanent commitment as a matter of policy.

However, it must be recognized that subsidies, regardless of how good the intentions of the government or development organization providing those subsidies, may hurt rather than promote the penetration of PV technologies.

4.1 How subsidies can hurt

The fact that temporary equipment and financial subsidies for PV can hurt has already been recognized (Desai (Ref.29):

"Heavy subsidies given to selected alternative energy after the oil crises by developing countries have led to either little energy market penetration or have been of doubtful economic value. Subsidies to producers have created inefficient and fragile industries. Subsidies to products have frozen technology."

And there are unfortunately too many cases history to verify this observation:

From the Asia Alternative Energy Unit of the World Bank (Cabral (Ref. 30)):

"A study of 37 rural electrification programs reveals that most programs were projected to cover only 50-80% of total costs"

"Recovering the capital investment, as well as the cost of borrowing, administration and maintenance are crucial to the financial sustainability of any SHS programs."

"Assuming that rural customers have a very low capacity to pay, Government sponsored ... programs set low monthly fees, based on household kerosene expenses. Such programs are intrinsically not sustainable and often result in poor cost recovery."

From the Sri Lanka Pansiyagama project (Ref. 31):

"...beneficiaries considered the program a Government give-away. Cost recovery was only 52 percent, even though a minimum payment of $1.5 to $1.5/month was required. "

On the Panpres project, in Indonesia:

"The average cost recovery ($3.75/household/month)...was about 60 percent although it ranged from 25 to 95 percent among participating cooperatives. Several subsidized schemes in the Pacific Islands have failed due to financially unsustainable pricing policies."

On the Tuvalu project:

"the life-cycle cost of systems for a rural household lighting system are $1,386 over al 5 year period at net discount rate of 10%. Of that amount, the initial equipment costs are $873, so the value of and periodic replacement of batteries, etc., is $1386 - $873 = $513. Using the same time period and discount rate, the value of revenue collections is $465. Therefore, collections are insufficient to cover the initial equipment costs and the result is donor dependency."

These are just a few examples of PV subsidized projects to which the word "sustainable" cannot be applied in an economic sense. There is nothing wrong with the technology, or with the intentions of the donors. Is there anything missing in our understanding of the subsidies?

4.2 Debunking the myths

Once the environmental and social advantages of PV systems are recognized, there are many reasons why governments and international or private organizations decide to provide subsidies for PV systems. Some of these "reasons" are rather like myths, and it may be of interest to review some of them.

Myth 1. *"People cannot afford PV systems."*

This myth results partly from a mixture of fact and the mixing of perspectives. Unquestionably, capital costs for PV are high, and the PV installation for a home using as much energy as typical American or European homes use is costly. Unquestionably, many people in the rural countries of the developing world are very poor indeed. So?

How much are those people in the rural areas of the developing world spending on kerosene, dry cells, and automotive batteries? If PV sources can provide "services" (light or electricity for radio sets, for example) at a lower cost than the alternatives now being used, what does it mean to state that "people cannot afford PV"?

Myth 2. *"People generally need larger systems than they can afford, even with financing."*

Yes, and most of us "need" more things than we can afford, even with financing. Most people can be satisfied with a smaller system than they supposedly need. Most of the time, a small system is better than no system at all - and PV systems can always be expanded later.

Myth 3. *"Subsidies are necessary to introduce the technology because of the risk."* This is a costly approach, and one that fails to communicate to the true value of the systems. Experience shows that it may be easier to introduce the technology by charging full price than to attempt full cost recovery *after* people have grown used to subsidies.

If the perception of risk is an obstacle to market penetration, would it not make more sense to make provision for trial periods and guarantees?

Myth 4. *"A subsidized project will demonstrate acceptance and viability, and will lead to private investment."*

Full cost recovery models can accomplish the same goal, with an important difference affecting sustainability. Investors may avoid areas with a history of equipment subsidies, unless there is reasonable assurance that the subsidies will continue indefinitely.

Myth 5. *"Subsidies enable a more equitable distribution of resources."*

There are large income variations in all rural areas of the world, and a fixed subsidy to all beneficiaries does not promote equity. Those with more resources should be allowed to buy or lease their own systems; assistance to the economically disadvantaged should be provided as a form of social safety net to the extent that funds are available for such purposes.

Myth 6. *"It is inappropriate to charge fees for donated project components."*

There are practical - and often modest - limits to how many components can be donated. Rather than offering a one-time equipment give-away, it would make sense to use donations of money or components to create a capital base that reduces the need for future grants. Pricing strategies that cover equipment depreciation and all other costs enable the leveraging of resources in mainstream capital markets for project expansion.

4.3 The Real Effects of Unnecessary Subsidies

What is the real effect of the typical intermittent equipment subsidies?

Donor Dependency

Clearly, a generous enough subsidy may make it possible for a donor or a government to show quick results in terms of the number of installations, but the installations are rarely if ever sustainable. Instead, subsidies create a *dependency on donors*. About a PV experience in a Pacific Island, Liebenthal, Mathur and Ade (Ref. 31) write:

"TSECS earned roughly A$1 per month, a level of tariffs that provides for operation and maintenance costs but not for the expansion or the replacement of the solar panels at the end of their useful life. Consequently, at present TSECS does not have the capital to provide new installations and must rely on donors."

Waste of scarce resources:

Money spent on equipment subsidies may be invested in other activities which are more likely to result in the sustainable use of the technology, such as the establishment of credit pools, technical assistance training and project operation.

We should all remember what good pavement good intentions make. Many people with good intentions seem to forget that most human beings are more likely to maintain and care for something in which they make a personal investment.

Weakening the market

A big threat to the small legitimate commercial supplier in a developing country is that at any time a large enough donor may compete unfairly. There are other more subtle ways in which donations and well-intentioned subsidized projects can weaken the market, however. For example, donors may place large orders, requiring local business to gear up for a demand which is artificial and cannot be sustained.

What should developers and planners do?

Developers and Planners should adopt pricing strategies that reflect market conditions.

They should understand that it is possible to recover fully investments in PV projects, even in relatively poor countries. There is an old Chinese proverb about how much better than given away fish to the poor is to teach the poor how to fish. Developers and planners should aim at assisting project developers and funding sources to catalyze financially sustainable projects.

There are examples of successful sustainable projects, and the authors of this paper can share their own experience and that of others, drawn from Enersol Associates, Inc., a nonprofit international development organization specializing in PV project implementation, and from SOLUZ, Inc., A Massachusetts-based company affiliated with Industria Eléctrica Bellavista, which operates a PV system leasing business in the Dominican Republic. Together, these organizations have facilitated the installation of over 5000 systems on a full cost recovery basis.

Similar success stories are being repeated elsewhere in the world: in India, in Sri Lanka, etc. The key to sustainability is full cost recovery.

5. FULL COST RECOVERY

What is Full Cost Recovery? Ideally, the goal that a project allows "full cost recovery" means that the burden of all costs are borne by the end-user, signaling the readiness of a project to survive financially without subsidies from tax payers or donor organizations.

This "ideal" must be interpreted judiciously. Every option (coal, oil, nuclear, hydro, kerosene) is somehow subsidized, and there is really no such thing as a "level playing field". If PV rural electrification, it makes sense to subsidize solar programs to level the field.

We have seen that subsidies can do more harm than good: they reinforce the stigma that weakens the solar market and turns solar energy into the "energy source for the future, where it will always be. The question is, what is an appropriate subsidy?

Projects should strive to demonstrate earnings potential commensurate with the level of investment risk. Given the current trend towards privatization and the global experience with small, stand-alone PV systems, a key criterion that should be applied to determine the appropriateness of a subsidy is its effectiveness in creating conditions conducive to private investments in photovoltaic projects. To make this determination, it is helpful to recognize that there are many cost components to any photovoltaic development program, and that changes in the different cost items may have very different impact on the sustainability of the program.

The costs associated with a photovoltaic development program may be categorized as program development costs; equipment, operation and maintenance costs; and financing costs, as follows:

Program costs

Program Development Cost
Planning
Training
Technical Assistance
Policy making

Equipment, O&M costs
Equipment (depreciation)
Transportation
Taxes and duties
Wholesale margins

Installation margins
Warrantee
Administration

Financing Costs
Interest
Contracting
Collection
Fund administration

Which subsidies hurt? *In general, projects with substantial equipment, operation and maintenance, or financing subsidies do not help create conditions conducive to private investments.*

There are several reasons why these projects are generally not sustainable:
1) they collect insufficient revenues to demonstrate earnings potential to investors
2) they create expectations on the part of potential paying customers that photovoltaic systems can be obtained for below market prices, and
3) they misinform the investors about people's ability to pay for the technology, and hence, about the real risks of investing in projects.

A project in an area with a history of equipment subsidization will lure private investors only if the subsidies can be expected to continue long into the future - but that is not generally the stated, long-term intention of most PV development projects.

On the other hand, training and technical assistance, as well as some other forms of "market conditioning" or project development, do create conditions conducive to private investment. This is because such assistance and conditioning lower the investment risks. Furthermore, the need for external training and technical assistance may (and should) decrease over time, with little or no effect on the price of photovoltaic systems.

Loan guarantees and some forms of direct public financing also attract private investment, and are acceptable subsidies because they are common in many "commercial" financial transactions (e.g. Ex-Im Bank) and are likely to continue as such.

Donated equipment can also support the goal of eventual investments in PV projects, provided that the end-users are charged market rates for the equipment and/or services they receive. Such a donation can be helpful if it contributes to a pool of equity that can actually leverage other financing.

In summary, project developers should strive to recover, at the very least, all equipment, operation and maintenance and financing costs that cannot be expected to be subsidized well into the future.

6. FINANCIAL ARRANGEMENTS FOR FULL COST RECOVERY

There is a wide range in the financing and institutional options to supply solar electricity to rural areas. Full cost recovery may be achieved through cash sales, consumer loans and lease assistance. It may be useful to outline these options.

6.1 "Cash" sales

(The end user owns and self finances.)

There is a market for cash purchases of photovoltaic system in the rural areas of the developing world. Half of all systems installed in the Dominican Republic and Honduras over 20,000 systems installed in Kenya have been bought outright by the end users. It is easy to overlook the size of the market: more than 50% of the over 300,000 solar systems installed world-wide have been paid in cash or on a very short-term (i.e., two to three months) basis.

This option "skims the cream" of the relatively affluent users. Although the market is obviously limited, it is huge. Other factors to note in this approach:

- Cash sales are "practical." They tend to occur at a slower pace than do financed installations, and so they allow newly trained technicians to gain experience.
- Even if financing is available, cash sales are important. In full cost recovery projects, many of the roughly five percent of the rural unelectrified population that can buy systems outright will do so. This frees scarce capital to finance other schemes.
- Businesses that sell to cash customers understand the concept of full cost recovery. They must charge for all costs. In competitive situations, private businesses generally operate with very modest overhead and profit margins.

People who purchase PV generally must pay for maintenance and repairs beyond the warrantee. Some analysts argue that schemes that build in regular maintenance routines are necessary, but Enersol's experience does not support this argument: cash customers can usually afford the upkeep and will not allow their investment to remain idle.

6.2 Consumer Loans

(User purchases the system with term financing from equipment vendor (conditional sale), non-government organization or bank.)

Vendor Financing

A significant portion of PV installations are financed directly by the vendors, usually for one to three months at market or slightly above market rates.

Currency exchange variation introduces uncertainty, and rates may be stated in U.S. dollars to avoid devaluation risks. However, few vendors have sufficient capital or administrative capacity to offer medium- or long-term credit. The costs of vendor financing are those of capital, collection, currency devaluation, default, overhead, and often deferred profit.

In Honduras and the DR, vendor-financed systems account for an estimated 25 percent of all installations.

Revolving Funds

A revolving credit fund may be initiated with donated capital. Terms have ranged form one to three years with interest rates within a few percentage points above prevailing commercial rates.

The fund managers (usually community development associations, agricultural coops and the like) pay the equipment vendor/installer for those services not covered by the down payment, and they receive periodic payments from the customers.

The end-user owns his or her own system and pays for maintenance as required (as with cash purchases), but the system serves as collateral until the loan is paid off. Finance costs include those for vendor financing, with additional costs of marketing of the financial product and loan processing. It is prudent to assume that the next cash infusion will come form wholesale-rate loan, even though the fund may have been capitalized with donations.

Managing a revolving fund requires a level of expertise that is not usually found at the village level.

A key element of cost recovery involves professional management or accountability to an outside organization. It is not usually realistic to expect full cost recovery if funds are simply donated to village level or informal organizations in the absence of external managerial inputs.

Commercial Banks - Loan guarantees

Commercial banks have not commonly given credit for single home PV systems, and those that do so require salaried cosigners or mortgage guarantees that few rural people have.

Also, location is a problem: the users often live far from the cities and towns where the commercial banks are.

It is possible to solve this problem. Enersol has done so through its Fondo Solar, a fund which provides loan guarantees to non government organizations. With an initial grant from the Rockefeller Foundation, Enersol deposited $85K in a U.S. interest-bearing account in 1993, and later borrowed $30K from the IFREE. Using that fund as collateral, guarantees are provided to local banks in Honduras and the Dominican Republic for project loans, shielding the banks from most risks. Enersol is protected from currency related risks because the local banks, not Enersol, issue loans, and the banks have generally given preferential interest rates based on this guarantee.

It is possible to guarantee local bank retail loans to hundreds of end-users, or to guarantee "wholesale" loans to agencies that lend to the systems buyers.

Between 15 and 25% of the unelectrified population in Honduras and the Dominican Republic (between 20,000 and 200,000 homes) could afford systems financed in three years. With an investment of $1 million, the Fondo Solar could finance 8000 systems in five years.

6.3 Lease Assistance

(End user leases from equipment supplier, utility, non government organization, or other entities.)

Many rural households are willing to pay a monthly lease (or energy service) fee. Two schemes for leasing remote home and business systems are possible:

- Users may pay a month fee for several years until they effectively pay for their systems, at which time they assume ownership of the systems, or
- The lessor simply sells the energy service and never relinquishes ownership.

Because the financing is long term, a greater percentage of the population can afford the monthly payments. Ultimately, the users spend more money over time on leasing fees than they would by simply purchasing an equivalent system, but they are not required to accumulate capital for a down payment.

The cost for a system leasing are a combination of types of costs incurred by equipment vendors and financiers. A significant level of organizational sophistication is required to manage a leasing operation efficiently. Soluz, Inc. in the Dominican Republic provides one developing country example in which full cost recovery is expected from a leasing project.

The arrangements outlined above require different disbursement schedules from the end users. Which one is adopted will depend on the local conditions, financial climate. To some extent, they may well depend on the preference of the local government, and this preference may be affected by local history, social structure, and even ideology. If the authorities believe that individual ownership is an important factor for the success of the program, this will certainly affect the type of arrangement that are supported.

The emphasis of this paper, however, is on sustainability and full-cost recovery, and in principle any of the arrangement just outlined is compatible with this emphasis.

CONCLUSION

Decentralized solar photovoltaic systems can help meet global energy demand in the next decades, and has environmental and social advantages. Energy and economic development are closely intertwined: as productivity increases through electrification, more resources become available, and those resources expand the base of those for which the technology is affordable.

This virtuous cycle is likely to be stable, but there is a need for an effective pioneering effort that makes the technology acceptable by nurturing the organization of the needed infrastructure and identifying the human resources as well as the needs and limitations of the market. Such a pioneering role is one which has been played by Enersol Associates, and this is the role that we suggest must be played anywhere if the market penetration is to be sustainable. Efforts such as Enersol's open up the communities to the acceptance of the technology, sharing the benefits and the responsibilities with community organizations, empowering the communities to make choices, identifying technicians, and setting up the ground for a rural infrastructure for the installation and servicing of systems.

In the early stages the "pioneers" need to explore the territory, build an experience base and an institutional track record and prepare the groundwork for a more substantial scale up. As the process continues, the pioneer or their successors evaluate the capacity that is present and make determinations as to the rate of expansion and the local capabilities. When the time and the conditions are right, sector activities can expand based upon the successful preliminary human and institutional capacity building activities and the sustainable application of the technology.

Successful examples of solar rural electrification projects around the world have been shown that it is possible to reach very poor elements of society with private initiative and investment, and it is rational to adopt development policies that encourage the market penetration of those systems through subsidies.

However, a key criterion that should be applied to determine the appropriateness of a subsidy is its effectiveness in creating conditions conducive to private investments in photovoltaic projects.

Development assistance should focus on risk reduction through planning, technical assistance, training, and policy making

Another appropriate role for donors and development assistance agencies is to provide seed capital. Grant or concessional "equity" should be used to leverage further borrowing from local capital markets whenever possible.

Project developers and planners should adopt pricing strategies that reflect market conditions.

Trade and industry promoters should resist the temptation to promote "parachute" approaches to technology dissemination and instead support initiatives that cultivate a broader demand for the product over time.

Project planners and developers should encourage competition between local equipment supplies that charge full costs for good and services provided directly to end-users, including the marginal costs of capital, equipment and O& M.

Whereas temporary equipment and financial subsidies are not helpful for securing private resources over the long term, an emphasis on risk reduction in the form of initial infusions of project development support for efforts that feature full cost recovery can attract capital to make the photovoltaic technology broadly available.

This will benefit rural people in developing countries - and help alleviate the formidable environmental and financial challenges facing all of us in the next few decades.

REFERENCES

1. Dörner, D., "Transformação de consciência ecológica,", Deutschland, Special Portuguese version, Societäts-Verlag, ISSN 0945-683X, Frankfurt, Germany (1995)
2. Chen, Robert S., "Environmental Risks of Energy Production: The Carbon Dioxide Example," A Global View of Energy, Kursunoglu, Millunzi and Perlmutter, editors, Lexington Books, 1982
3. Carr, Edward, "Power to the people," The Economist, June 18th, 1994

4. Arthur, Edward, "Accelerator System for Plutonium Destruction and Waste Transmutation," Proc. of Global Conference on Energy in Transition - (Global Conference), Plenum Press (in press)
5. German, Michael, "Natural Gas and Electricity," Proc. of Global Conference on Energy in Transition, op. cit.
6. Schleede, Glenn, "An Objective Analysis for Gas-Fired Electricity Generation in the U.S.," Global Conference, op. cit.
7. Moore, Taylor, "Developing Countries on a Power Drive," Electric Power Research Institute Journal, Volume 20, Number 4 (1995)
8. Garrity, Thomas, "US/World Electric Generation Forecast," Global Conference, op. cit.
9. Hande, Harish, and Martín, José, "Photovoltaics for Rural Electrification in India," Proc. of Solar Energy Society Conference, Quebec, 1993.
10. Kursunoglu, Behram, "Foreseeable Expansion of the Global Market for Electricity," Global Conference, op. cit.
11. Brown, Lester R., "Analyzing the Demographic Trap," State of the World, W. W. Norton & Co., N. Y. 1987
12. Leal, Ondina Fachel, Department of Anthropology, UFRGS, private communication (1994)
13. Starr, Chauncey, "Global Energy and Electricity Futures," Global Energy Conference, op. cit.
14. Jones, Donald W., "How Urbanization Affects Energy-Use in Developing Countries," Energy Policy, September 1991.
15. Martin, J. and Hande, H., "From Bellavista to India: Extrapolating by Three Orders of Magnitude," Proc. of Global Forum on Energy and the Environment, New Delhi (1993)
16. Hande, H., Wijesooriya, Hansen, R., and Martin, J. "On Why Solar Energy Makes Sense in the Developing World," Proc. of 12th European Photovoltaic Conference, Amsterdam (1994)
17. Blitzer, C. R., Eckaus, Lahiri, S. and Meeraus, A., "How Restricting Carbon Dioxide Emissions Would Affect the Indian Economy," Working Papers for the World Bank World Development Report, 1991
18. World Development Report, World Bank, 1992.
19. Pearce, David, Adger, Neil, Maddison, David, and Moran, Dominic, "Debt and the Environment," Scientific American, Vol. 272, 1995, p. 52
20. Teller, E., "A Fool-Proof and Dictator Proof Fission Reactor," Proc. of Global Conference on Energy Demand in Transition, op. cit.
21. Rosenfeld, Arthur, The Transition from Fossil Fuel to SE - North America," Global Conference, op. cit.
22. Steven Wiel, "The Transition from Fossil Fuel to Sustainable Energy - Developing Countries," Global Conference, op. cit.
23. Hansen, R., and Martin, J., "Photovoltaics for Rural Electrification in the Dominican Republic," Natural Resources Forum, New York, 1988.
24. Gregory, R. Flynn, J. and Slovic, P. "Technological stigma," American Scientist, Vol. 83 (1995), p. 220.
25. Lysen, Erick, JH., "Photovoltaics in the South," Netherlands Agency for Energy and the Environment, Utrecht, 1994
26. Smith , Frank, "Policy Reform and Sustainable Energy Development, " Consultation on Renewable Energy and Sustainable Development in Latin America and the Caribbean (1994)
27. Silverman, Murray, and Worthman, Susan, (San Francisco State University) "The Future of Renewable Energy Industries," The Electricity Journal, 8, No. 2, March 1995, p. 12.
28. Lazonick, William, - "Business organization and the myth of the market economy," Cambridge University Press, NY, 1993
29. A. V. Desai, "Alternative energy in the Third World: a reappraisal of subsidies," World Development, July 1992 v. 20, N 7 p 959 (7)
30. Cabraal, Anil and Cosgroves-Davies, Malcom, "Solar Photovoltaics: Best Practices for Household Electrification," The Asia Alternative Energy Unit, the World Bank, Washington DC, October, 1994.

Chapter III

The Next Fifty Years of the Peaceful Application of Nuclear Energy

THE GAS TURBINE–MODULAR HELIUM REACTOR FOR THE NEXT FIFTY YEARS OF NUCLEAR POWER

L. S. Blue

General Atomics
P.O. Box 85608
San Diego, CA 92186-9784

ABSTRACT

The Gas Turbine-Modular Helium Reactor (GT-MHR) is the result of coupling the evolution of a low power density passively safe modular reactor with key technology developments in the U.S. during the last decade: large industrial gas turbines; large active magnetic bearings; and compact, highly effective plate-fin heat exchangers. This is accomplished through the unique use of the Brayton cycle to produce electricity with the helium as primary coolant from the reactor directly driving the gas turbine electrical generator. This cycle can achieve a high net efficiency in the range of 45% to 48%.

In the design of the GT-MHR the desirable inherent characteristics of the inert helium coolant, graphite core, and the coated fuel particles are supplemented with specific design features such as passive heat removal to achieve the safety objective of not disturbing the normal day-to-day activities of the public even for beyond design basis rare accidents.

Each GT-MHR plant consists of four modules. The GT-MHR module components are contained within steel pressure vessels: a reactor vessel, a power conversion vessel, and a connecting cross vessel. All vessels are sited underground in a concrete silo, which serves as an independent vented low pressure containment structure.

By capitalizing on industrial and aerospace gas turbine development, highly effective heat exchanger designs, and inherent gas cooled reactor temperature characteristics, the passively safe GT-MHR provides a sound technical, monetary, and environmental basis for new nuclear power generating capacity.

This paper provides an update on the status of the design, which has been under development on the US-DOE program since February 1993. An assessment of plant performance and safety is also included.

INTRODUCTION

The challenge for the nuclear industry in the 1990s will be the development of plants that address matters of public acceptance, improved safety, competitive generating costs and reduced radioactive waste. The GT-MHR which has evolved from the steam cycle Modular High Temperature Gas Cooled Reactor (MHTGR) firmly addresses each of these issues.

Like the MHTGR, the GT-MHR relies on design selections and passive safety systems to assure retention of radionuclides within the TRISO coated fuel particles. Utilizing a special annular core design which is larger and located nearer to the reactor vessel and Reactor Cavity Cooling System than the MHTGR, peak fuel temperatures during a rare loss of forced cooling and pressure event are within the design goal for the fuel of 1600°C.

The GT-MHR is an environmentally acceptable power plant that has a high degree of inherent safety characteristics. These inherent safety characteristics coupled with a direct cycle gas turbine allow design simplifications that help control costs and schedule. The GT-MHR can uniquely use the Brayton cycle to produce electricity by directly driving a gas turbine electrical generator. The GT-MHR can achieve a net efficiency in the range of 45% to 48%. This high efficiency leads to competitive economics when compared to the 32% net efficiencies achieved by advanced water cooled reactor power plants. High thermal efficiencies lead to an environmentally compatible design that produces about 50% less high level radioactive waste and about 100% less thermal discharge to the environment than comparably sized LWRs (Ref. 1).

The GT-MHR evolved from a 10 year design effort on the steam cycle MHTGR funded under the U.S. DOE Advanced Reactor Program. Producing ~286 MW(e) per reactor module, the GT-MHR retains many of the key MHTGR design features including refractory coated TRISO fuel, low power density core, factory fabricated steel vessels, below grade siting, and completely passive decay heat removal. Most importantly the GT-MHR retains the capability of meeting all safety goals without relying on active safety systems or operator actions. The GT-MHR couples this impressive safety performance to several key technology developments of the last decade: large industrial gas turbines, large active magnetic bearings, and compact plate-fin heat exchangers. The major difference from the steam cycle design is instead of using steam to drive a steam turbine plant at 38% of efficiency, the GT-MHR produces electricity directly in a closed-cycle helium turbomachine at 45% to 48% net efficiency, thereby eliminating the expense and complication of the steam plant equipment.

The GT-MHR addresses the issues of the 1990s by providing a step increase in economic performance combined with reduced environmental impact and increased reactor safety.

GT-MHR PLANT DESIGN

In the design of the GT-MHR the desirable inherent characteristics of the inert helium coolant, graphite core, and the coated fuel particles are supplemented with specific design features to ensure passive safety. Radionuclides are essentially retained under all licensing basis events within the refractory coated fuel particles. The integrity of the particle coatings as a barrier is maintained by limiting heat generation, assuring means of heat removal, and by limiting the potential effect of air and water ingress under potential accident conditions. The design of the GT-MHR provides redundant and diverse active systems to perform these functions for both normal and transient conditions. However, consistent with the safety and performance objectives, the fuel integrity is maintained because of inherent MHR characteristics and passive design features without the need for active AC powered systems or operator action.

The key features and design selections to ensure the GT-MHR's safety goals include:

- *Helium Coolant* - The inert and single-phase helium coolant has several advantages. No flashing or boiling of coolant is possible, pressure measurements are certain, no coolant level measurements are required, and pump cavitation cannot occur. Further, there are no neutronic reactions with the helium and no chemical or energetic reactions between coolant and fuel.

- *Graphite Core* - The strength and stability of the graphite core and the ceramic fuel at high temperatures result in a wide margin between either operating temperatures or accident temperatures and temperatures that would result in core damage. Further, the low power density of the core and the massive graphite core structure with its large heat capacity ensure that changes in the overall core temperature take place very slowly.
- *Coated Fuel Particle* - The coated fuel particles consist of microspheres of uranium oxycarbide kernels clad with layers of pyrolytic carbon and silicon carbide. This TRISO coating is stable under irradiation and prevents significant release of radionuclides for long times (several hundred hours) even at temperatures reached in severe accidents. The 770 mm diameter coated fuel particles are bonded into rod-shaped fuel particle compacts with a graphitic binder and inserted into cylindrical holes, drilled in the hexagonal graphite fuel elements.
- *Negative Temperature Coefficient of Reactivity* - The nuclear characteristics of the graphite and low enriched uranium materials combine to produce a power coefficient dominated by the temperature coefficient of reactivity, which is strongly negative for all operating and accident conditions. This large negative temperature coefficient will terminate the nuclear reaction if the core heats to beyond normal operating temperatures.
- *Core Power and Power Density* - The maximum thermal output of the reactor core and the core power density have been kept low to limit the amount of decay heat that must be dissipated during an accident - a key factor in enabling passive safety.
- *Core Geometry, Reactor System* - The annular geometry, large height to diameter ratio, and the uninsulated steel reactor vessel and redundant, passive reactor cavity cooling system have been elected to ensure adequate decay heat removal from the core through passive thermal radiation, conduction, and natural convection.

The use of inert helium gas coolant with a graphite core and TRISO coated fuel facilitate operation of the reactor at temperatures compatible with the safe generation of electricity with high thermal efficiency.

The GT-MHR module arrangement is shown in Figure 1. Each GT-MHR plant is envisioned to consist of four modules. The GT-MHR module components are contained within three steel pressure vessels: a reactor vessel, a power conversion vessel, and a connecting cross vessel. All three vessels, which are made from high strength 9Cr-1Mo-V alloy steel, are sited underground in a concrete silo that serves as an independent vented low pressure containment structure.

The 8.4m diameter by 31.4m high reactor vessel contains the annular reactor core, core supports, control rod drives, refueling access penetrations, and a shutdown cooling system. The reactor core is formed by hexagonal graphite fuel columns, which contain a mixture of 20% enriched fissile and natural uranium fertile fuel encapsulated in ceramic coated microspheres.

The reactor vessel is surrounded by a reactor cavity cooling system (RCCS), which provides totally passive decay heat removal. The separate shutdown cooling system provides backup decay heat removal for refueling and maintenance activities.

The power conversion vessel contains the entire power conversion system as shown in

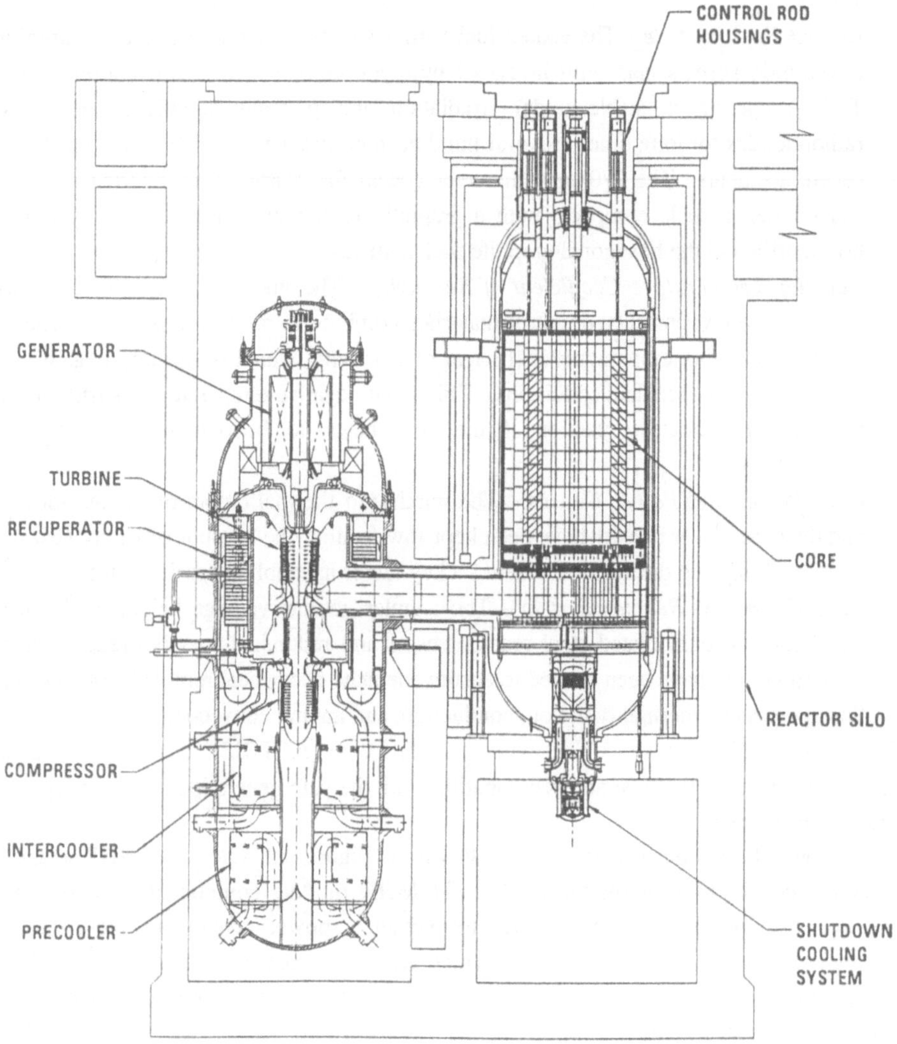

Figure 1. GT-MHR Plant Arrangement

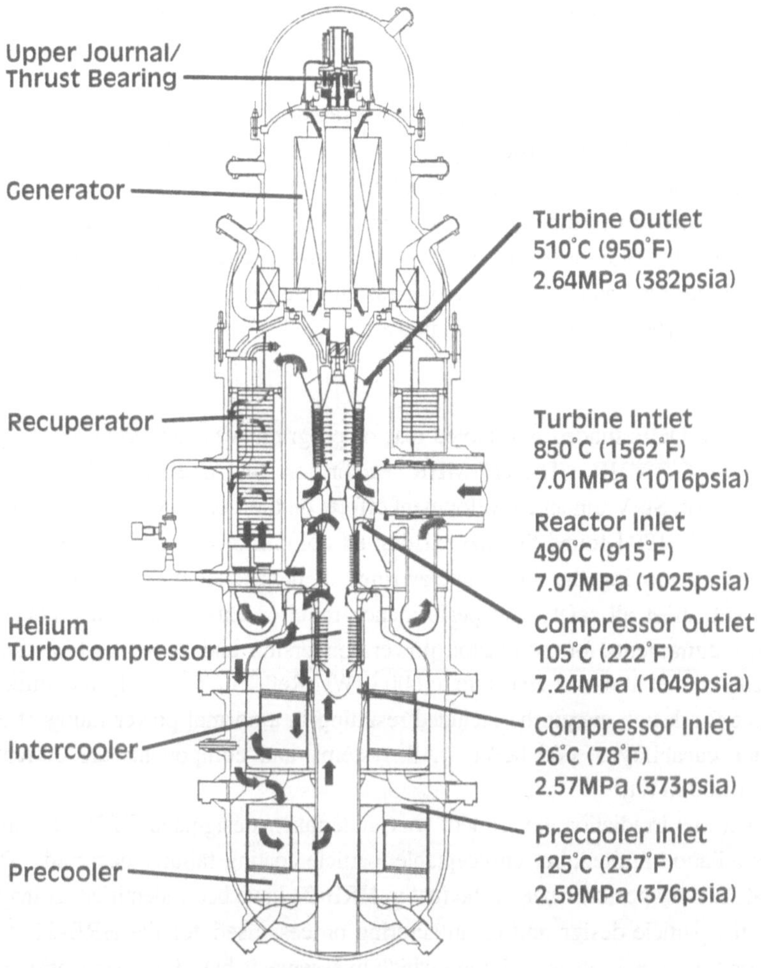

Figure 2. GT-MHR Power Convertion Vessel Assembly

Figure 2. The turbomachine consists of a generator, turbine, and two compressor sections submerged in helium, mounted on a single shaft supported by magnetic bearings. The power conversion vessel also contains three compact heat exchangers: the high efficiency plate-fin recuperator and the water cooled intercooler and precooler.

The major component in this vessel is the vertical single-shaft helium gas turbine that drives the synchronous generator. The gas turbine consists of two separated compressor sections which permit intercooling and the turbine. The rotor assembly uses active magnetic thrust and journal bearings. The GT-MHR gas turbine is physically smaller than gas turbines in service due to the high coolant pressurization associated with closed cycle systems.

The three heat exchangers located within the power conversion vessel contribute significantly to the plant's high efficiency. The turbine exhaust is used by the compact plate-fin recuperators to preheat the high pressure compressor discharge helium before it enters the reactor resulting in enhanced efficiency. The low temperature duty helium-to-water heat exchangers provide low inlet temperature helium to the compressors. These helical bundle heat exchangers use finned-tube geometries and operate in a far less demanding temperature environment in the Brayton cycle than in the Rankine cycle. Furthermore, these heat exchangers reduce the potential for water ingress into the helium coolant since the helium is at a higher pressure when the plant is at full power.

DESIGN STATUS

In 1993, the DOE funded gas-cooled reactor program was redirected to design studies focused on the GT-MHR. The GT-MHR reactor core and internals build on technology developed for Fort St. Vrain and development efforts on the steam cycle MHTGR. A power level evaluation in 1994 found that the existing 84-column reactor core could be enlarged to 102 columns by moving the three annular rings of hexagonal fuel elements closer to the vessel and still meet all safety and performance requirements. The study concluded that component requirements for the reactor, power conversion and vessel systems could, with development, be met at power levels up to 600 MWt (Ref. 2). The study recommended that an allowance for design margin be included resulting in a normal power rating of 550 MWt and a stretch capability of 600 MWt. All systems and components are currently being designed for 600 MWt.

In 1991 a fuel irradiation experiment was undertaken designated HRB-21. In the later stages of irradiation limited, but unacceptable, particle coating failures occurred. The causes of the unsatisfactory performance of the fuel in HRB-21 have been identified as inappropriate changes in the particle design and manufacturing process used for the HRB-21 design. A consensus on fixes has been agreed upon which in essence is based on using prior successful technology. Manufacturing process improvements are ongoing, and the next capsule is scheduled for irradiation in 1996.

The helium turbomachine is in an early stage of design but sufficient work has been done to confirm that the performance goals can be realized (Ref. 3) and that the technology for a near-term plant with a turbine inlet temperature of 850°C (1562°F) has a firm basis in the U.S. (Ref. 4). Engineering studies indicate that vertical turbocompressors and submerged generators supported on magnetic bearings are feasible. The recuperator design draws on already proven technology used in the design of similar recuperators for propulsion and industrial gas-turbine plants. In these applications, the number of load cycles and transient conditions are much more severe than in the GT-MHR application (Ref. 5).

Overall the conceptual design of the GT-MHR plant is well advanced. Although significant engineering effort is required, no show stoppers have been identified. Schedule assessments indicate that the first plant could be deployed within 10 years given adequate funding.

PLANT PERFORMANCE

The GT-MHR process flow is shown in Figure 3. The helium coolant exits the reactor core at 850°C (1562°F) and 6.91 MPa (1003 psia), flows through the center hot duct within the cross vessel, and is expanded through the turbine in the Power Conversion System. The turbine directly drives the electrical generator and the high and low pressure compressors. The helium exits the turbine at 510°C (950°F) and 2.56 MPa (371 psia), flows through the high efficiency plate-fin recuperator to return as much energy as possible to the cycle, and then flows through the precooler to reject heat to the ultimate heat sink. Cold helium at 26°C (78°F) enters the intercooled compressor where it is compressed to 7.08 MPa (1027 psia) at 107°C (224°F) and then passes through the recuperator where it is heated. Helium at 488°C (910°F) and 7.00 MPa (1015 psia) flows from the recuperator exit, through the outer annulus within the cross vessel, past the reactor vessel walls for vessel cooling, and finally down through the core to complete the coolant loop.

The GT-MHR with a turbogenerator directly coupled with the reactor produces electricity with up to 45% to 48% net plant efficiency. An efficiency range is given which represents current estimates of how well leakages can be controlled within the Power Conversion

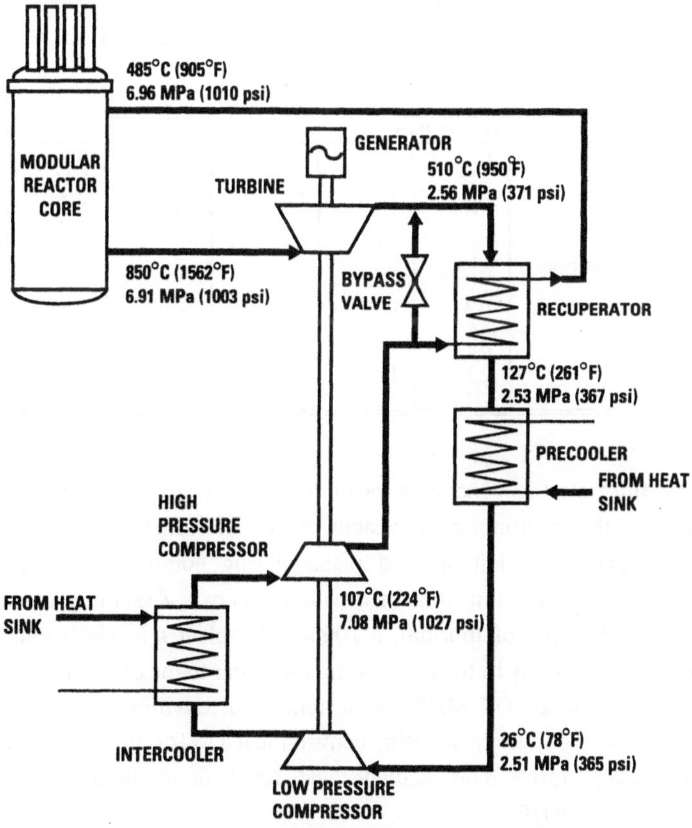

Figure 3. GT-MHR Process Flow Diagram

System. It also includes an allowance of 0.5% for adverse design evolution. Due to the high efficiency and consequent low thermal discharge, the GT-MHR has the flexibility to operate in dry environments by using dry cooling towers.

A preliminary economic evaluation has been performed for the GT-MHR using U.S. DOE guidelines for advanced reactors. The guidelines define a commercial or "target plant," for representing mature plant costs, as the plant which exceeds 4,500 MWe total installed capacity. All cost estimates are presented in terms of January 1992 dollars. Deployment schedules indicate that the target GT-MHR plant could begin power generation in approximately the year 2015.

The GT-MHR target plant performance parameters and economic evaluation results are compared to those of alternative mature plants in Table 1 (Ref. 6). Results are provided for two power levels, 550 MWt and 600 MWt per power unit. The mature GT-MHR is projected to have a lower busbar cost than any other alternative. The primary factors that account for the superior economics of the GT-MHR include: the high thermal efficiency of the closed Brayton cycle; the projection of stable nuclear fuel costs; and the simplicity and reliability of the gas turbine system, which provides for reduced capital cost, low O&M cost, and high operational capacity factors.

Table 1. Comparison of GT-MHR generation costs with power generation alternatives.

	GT-MHR Target Plant		ALWR 2X600	ALWR 1X1200	Coal IGCC	Gas CCCT
Performance Parameters						
Plant thermal rating, MWt	4x550	4x600	3659	3582	2625	2203
Net thermal efficiency, %	47.7	47.7	32.8	33.5	38.1	45.4
Plant electrical rating, MWe	1049.4	1144.8	1200	1200	4x250	4x250
Plant capacity factor, %	84.0	84.0	80.0	80.0	84.0	84.0
Plant Economic Evaluation Results, 1992 Dollars						
Total capital cost, M$	1740	1740	2034	1860	1611	531
Unit capital cost, $/kWe	1658	1520	1695	1550	1611	531
Total capital cost, $/kWe	33.9	31.2	54.5	45.5	48.4	11.2
Fuel cost, $/MBTU	1.31	1.27	0.77	0.77	1.45	2.33
Fuel real escalation, %/yr	0.0	0.0	0.0	0.0	1.0	2.2
Levelized busbar generation costs, mills/kWh						
- Capital	21.4	19.6	23.2	21.2	21.5	7.1
- O&M	4.6	4.2	7.8	6.5	6.6	1.5
- Fuel	9.4	9.1	8.0	7.9	18.5	38.4
- Decommissioning	0.7	0.6	0.9	0.9	0.1	0.0
Total busbar costs	36.1	33.6	39.9	36.4	46.6	47.0

Unlike fossil fired electric generating plants of similar size, the GT-MHR does not annually discharge to the environment thousands of tons of sulfur dioxide, carbon dioxide, and oxides of nitrogen. In addition, coal plants during combustion release significant amounts of the uranium and thorium contained within the coal. Assuming coal contains 1.3 ppm of uranium and 3.2 ppm of thorium, a 1000 MWe plant releases 5 tons of uranium (containing 70 lb of U-235) and 13 tons of thorium each year to the environment (Ref. 7).

The high efficiency of the GT-MHR has a number of environmental advantages. As shown in Figure 4, for each electrical kWh, conventional LWR's produce 50% more high level radioactive wastes, 150% more actinide, and 100% more thermal discharge to the environment than the GT-MHR.

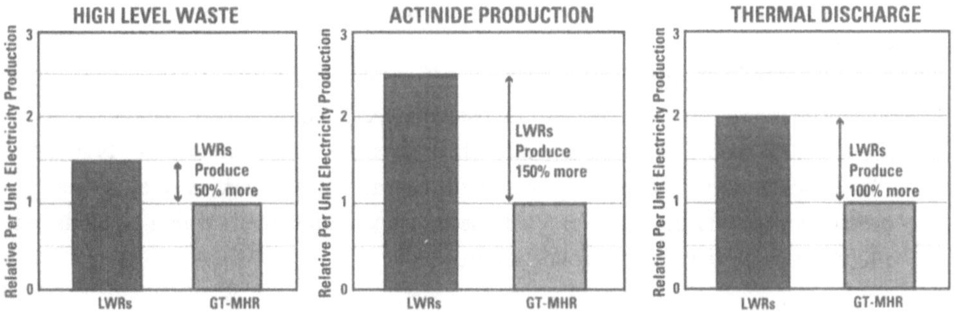

Figure 4. GT-MHR Minimizes Environmental Impact

SAFETY PERFORMANCE

The key to achieving large safety margins in the GT-MHR is in ensuring that the integrity of the fuel coatings, the initial barriers to radionuclide release, is not breached. Additional barriers such as the core graphite, vessel, and vented low pressure containment provide defense-in-depth to mitigate any releases and further increase the safety margins.

The fuel coating integrity is assured by controlling temperatures with ultimate reliance only on inherent characteristics and passive design features:

1. Control of Heat Generation

   Two independent active systems control reactor power levels. These are (1) the inner and outer control rod banks - a total of 54 control rods and (2) a diverse gravity-drop reserve shutdown system of 12 units. Similar systems were successfully demonstrated at Fort St. Vrain. The control rod system is fail-safe and inserts automatically on loss of power. Either system can maintain the core in a safe shutdown condition.

   In addition, a passive feature for heat generation control is provided by the inherent core negative temperature coefficient of reactivity. Following any interruption in forced cooling or a power excursion that increases core temperatures, this system mitigates the core temperature rise. Furthermore, even if forced cooling is completely lost and the active reactivity control system are unavailable, this feature alone will ensure reactor shutdown for more than one day. The German AVR reactor successfully demonstrated this inherent feature during a planned safety demonstration. Thus during an accident the reactor is shutdown, and only the core decay heat need to be removed to maintain acceptable core temperatures.

2. Decay Heat Removal

   Two active systems remove decay heat: the power conversion and the shutdown cooling systems. If neither active system is available, the independent passive reactor cooling system removes the heat. The core materials, power rating, power density, and configuration were chosen to ensure that sufficient heat could be removed to maintain acceptable core temperatures. Heat removal is accomplished by passive means through conduction, convection and radiation from the uninsulated reactor vessel to the passive reactor cavity cooling system. Thus, if neither active heat removal system is available, the continuously operating reactor cavity cooling system can reject heat to the environment by the natural circulation of air in an open loop system.

In the event that all active cooling systems fail and the vessel is breached such that the coolant pressure decreases to near atmospheric, peak nominal core temperatures of about 1521°C (2770°F) are reached after about three days in 5% of the core (Figure 5). The reactor cavity cooling system will cooldown the core without operator intervention or electrical power. At these temperatures fuel coating remains an effective first barrier to radionuclide release.

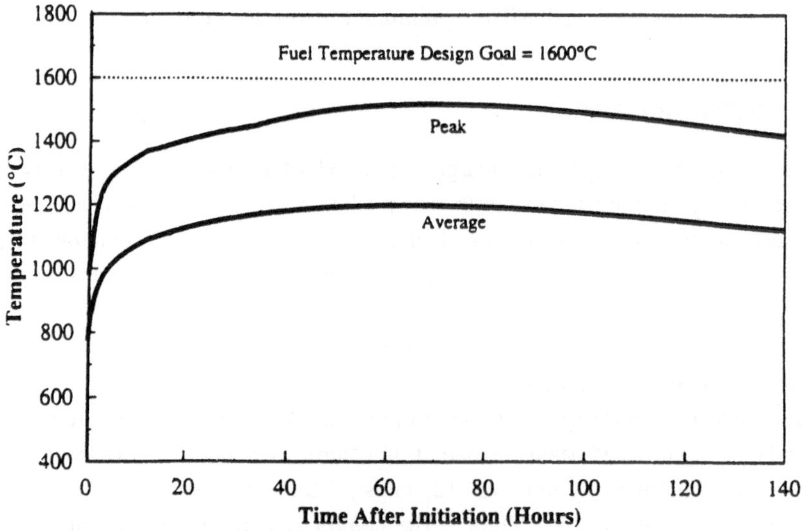

Figure 5. Average and peak fuel nominal temperatures during a 600 Mwt GT-MHR accident resulting in loss of all active cooling systems and loss of coolant pressure. Core cooling only on passive reactor cavity cooling system

An event in the GT-MHR that challenges core cooling is loss of a turbine blade (Ref. 8). The turbine deblading event results in a turbine and reactor trip and subsequent rapid internal helium pressure equalization in the primary coolant circuit. The event has the potential for disruption of the core geometry due to large pressure differentials or flow reversals across the core. Figure 6 shows the variation of coolant pressures at the core inlet and outlet plenums. Core inlet pressure remains greater than core outlet pressure, assuring that flow reversal is precluded. Analysis shows that primary system pressure decreases following the event from 7.00 MPa (1015 psi) to 4.1 MPa (60.0 psi) in 120s. During this time the pressure differential across the core increases to a peak of 0.13 MPa (19 psi) and then decreases as the flow decreases. The maximum pressure differential across the core is well below 0.96 MPa (140 psi) the maximum allowable compressive load. Therefore, core geometry is maintained, assuring a coolable core configuration.

3. Effect of Air and Water Ingress is Limited

Passive design features and inherent properties provide the basic defenses against chemical effects on the coated fuel particles. These include the choice of chemically

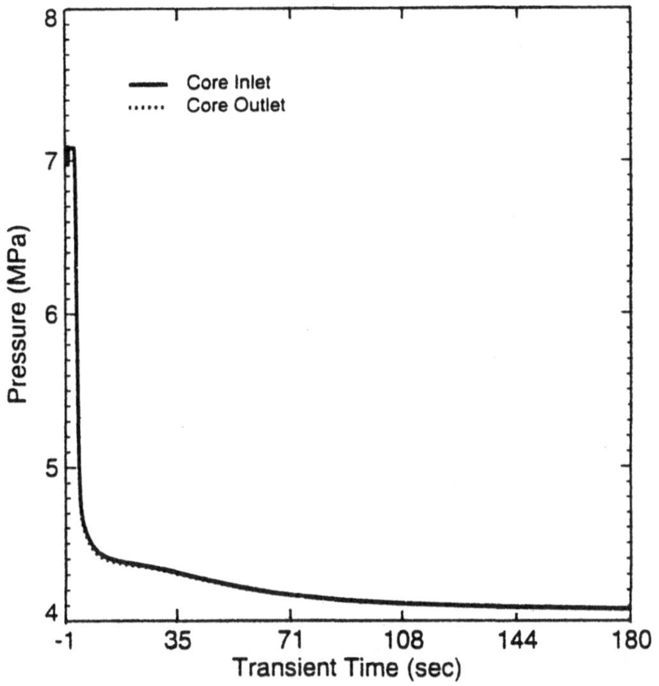

Figure 6. Transient pressure across the GT-MHR core during loss of turbine blade event

inert helium gas as the primary coolant to preclude any coolant reactions with the reactor or core and the use of multiple physical barriers between the particles and either air or water, the two chemical oxidants in close proximity to the core. Small air or water leaks into the primary system are quickly cleaned up by means of a continuously operating helium purification system. The effect on particles from even large amounts of air or water in the primary coolant is inherently limited by the massive graphite core and the coated particle, which itself is highly resistant to chemical effects from either.

Design features are incorporated in the GT-MHR to limit water ingress and its effect on the graphite core structure or the coated particles. These include limiting the number, pressure, and volume of water sources, locating the sources below the core elevation where possible, and providing a reliable water detection, isolation, and dump system. Of particular note is that the GT-MHR has eliminated the high pressure sources of water in the steam cycle steam generators, which significantly reduces the likelihood of water ingress during operation.

Even if the water enters the primary system, no significant oxidation of the graphite structure would occur unless temperatures were in excess of normal operating conditions. Further, the water graphite reaction is endothermic (absorbs heat) and is therefore inherently self limiting. If either of the two active cooling systems (primary and shutdown) are available, the core temperature can be maintained at or below normal temperatures. Even if the barriers are breached, releases of radionuclides will be limited to the inventory within the very small fraction of fuel particles which have failed silicon carbide coatings (from manufacturing defects or in-service failure).

The design features that limit air ingress are the use of nuclear grade vessels and the location of the vessels below-grade within the vented low pressure containment. With the arrangement of the primary system, multiple failures in the ASME Section III Class 1 vessels are required for air to displace the lighter helium. Even should vessel openings be available, rapid oxidation of the dense, high-purity nuclear-grade graphite by air requires temperatures well above normal operating temperatures. The temperature of the graphite can be controlled by either of the two active cooling systems. Should they not be available, the rate of any reaction is still limited because the air supply available to support the reaction is constrained by the high flow resistance of the coolant holes in the tall core structure. As a result, the amount of heat generated from the reaction of air with graphite would be small compared to the core afterheat, and radionuclide releases are within acceptable limits. Even if the air reacts sufficiently with the graphite fuel element structure and fuel matrix materials to reach the coated particles, the oxidation would be mitigated by the silicon carbide barrier layer on the particles. Under no circumstances could graphite "burning" be sustained.

CONCLUSION

The GT-MHR is the result of coupling a small, passively safe modular helium cooled reactor with large industrial gas turbines; active magnetic bearings; and highly effective plate-fin heat exchangers. The passively safe GT-MHR provides a sound technical, economic, and environmental basis to address the challenges the nuclear industry faces in the 1990s. The GT-MHR is unique in that it is the only reactor concept which can provide a step increase in economic performance combined with inherent safety features. This is accomplished through its utilization of the Brayton cycle to produce electricity directly with the high temperature primary helium coolant from the reactor directly driving the gas turbogenerator. Although significant effort is required to complete design and much detailed work remains, the design is largely based on existing technology. Design progress is constrained by funding but no show stoppers have been found. The busbar power costs of the GT-MHR are competitive with all other nuclear and fossil energy options in the U.S. The plant is able to achieve these favorable economics while maintaining high levels of passive safety that eliminate the possibility of core melt. Thermal discharge to the environment is low, and releases of harmful greenhouse gases are eliminated.

In conclusion, the GT-MHR offers one of the best opportunities of addressing the key nuclear industry issues of the 1990s of public acceptance, competitive costs, and reduced environmental impacts with an innovative, cost competitive, and passively safe nuclear option.

ACKNOWLEDGEMENT

This paper was sponsored by the United States Department of Energy under contract No. DE-AC03-89SF17885.

REFERENCES

1. Zgliczynski, J. B., F. A. Silady and A. J. Neylan, "The Gas Turbine - Modular Helium Reactor (GT-MHR), High Efficiency, Cost Competitive, Nuclear Energy for the Next Century," GA-A21610, Paper Presented at the International Topical Meeting on Advanced Reactor Safety, April 17-21, 1994, Pittsburgh, PA.

2. Silady, F. A., A. M. Baxter, T. D. Dunn, G. M. Baccaglini, and A. A. Schwartz, "Module Power Rating for GT-MHR Helium Gas Turbine Power Plant," GA-A21764, Paper Presented at the International Joint Power Generation Conference, October 3-5, 1994, Phoenix, AZ.
3. McDonald, C. F., R. J. Orlando, and G. M. Cotzas, "Helium Turbomachine Design for GT-MHR Power Plant," GA-A21720, Paper Presented at the International Joint Power Generation Conference, October 8-13, 1994, Phoenix, AZ.
4. McDonald, C.F., "Enabling Technologies for Nuclear Gas Turbine (GT-MHR) Power Conversion System," ASME Paper 94-GT-415.
5. McDonald, C. F., "GT-MHR Helium Gas Turbine Power Conversion System Design and Development," GA-A21617, Paper Presented at the American Power Conference, April 25-27, 1994, Chicago, IL.
6. LaBar, M. P., and W. A. Simon, "Comparative Economics of the GT-MHR and Power Generation Alternatives," GA-A21772, Paper Presented at the International Joint Power Generation Conference, October 3-5, 1994, Phoenix, AZ, USA.
7. Gabbard, Alex, "Coal Combustion: Nuclear Resource or Danger?" Oak Ridge National Laboratory Review, Vol. 26, No. 3 & 4, 1993.
8. T. D. Dunn, L. J. Lommers and V. E. Tangirala, "Preliminary Safety Evaluation of the Gas Turbine - Modular Helium Reactor (GT-MHR)," GA-A21633, Paper Presented at the International Topical Meeting on Advanced Reactor Safety, April 17-21, 1994, Pittsburgh, PA.

NUCLEAR POWER IN ITS SECOND HALF CENTURY

Shelby T. Brewer, Chairman

ABB Combustion Engineering Nuclear Power
Windsor, CT 06095

Sketching out a vision for nuclear power over the next half century is a highly subjective task. Any such sketch is bound to be impressionistic and fanciful. I will be only 105 years old at the end of this time window, and I fully intend to be around to see if my forecast came true.

THE FIRST 50 YEARS -- WHAT WENT WRONG

The pioneers in the 1950s and early 1960s envisioned a multi-decade or multi-century trajectory for nuclear power, here and abroad. The strategy took policy form in the 1954 Atomic Energy Act, and the 1962 and 1967 Reports to the President. This national strategy is schematized in Figure 1. In the broadest terms, a converter[1] (e.g. Light Water Reactor) economy would constitute the first civilian deployment of nuclear power, and over time, as the uranium resource base became depleted, the system would transition to a breeder reactor economy. The converter reactor spent fuel would be reprocessed with the residual uranium being recycled into converter reactors, while the residual plutonium would be used as start-up inventory for the breeder economy. Depleted uranium tails from the enrichment process (part of the converter fuel cycle) would be used as blanket material in the breeders to generate more plutonium for further breeder capacity additions. Wastes from reprocessing would be disposed of in an appropriate form and medium in a national waste system. In addition to capturing residual energy values,reprocessing would also render the waste into a more benign and manageable form. In this trajectory, nuclear power would provide an essentially inexhaustible electrical energy source.

This trajectory was derailed ,or rather interrupted, by the nuclear power downturn in the early 1970s--the abrupt truncation of new converter reactor plant orders and the cancellation of many of the plants in the construction backlog. In looking to the next fifty years, it is instructive to review how this elegant strategy became derailed. How and why did this happen? What went wrong? It is my basic thesis that the failure causes were not technological, but rather institutional, financial, and political.

[1] I use the term "converter reactor" here to avoid slighting the HTGR in favor of LWRs. The LWR, more particularly the PWR, dominated the first deployment phase world-wide, largely because of the spin-off from the US Naval Reactors Program in the 1950s.

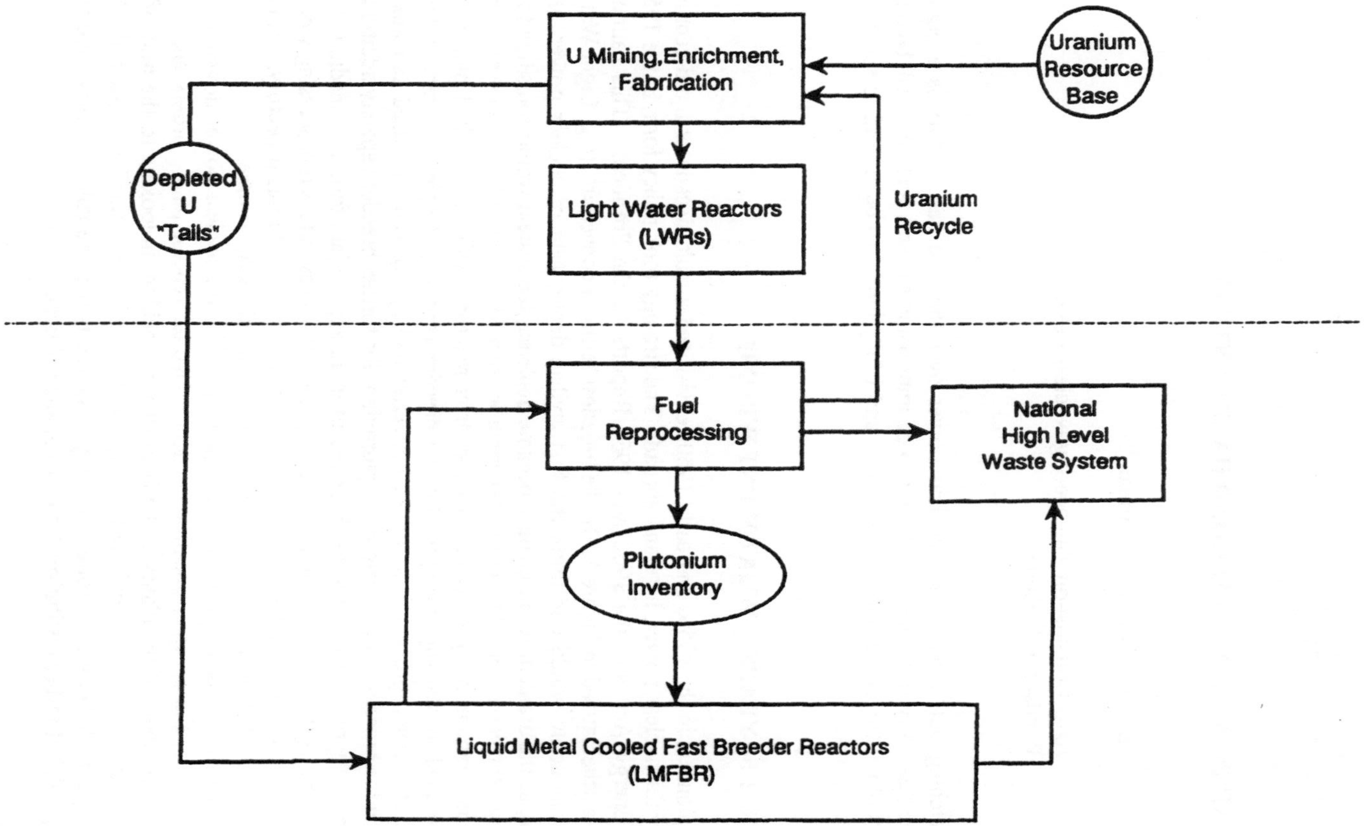

Figure 1. U.S. Nuclear Development Strategy: 1950s–1960s.

Several underlying forces converged in a narrow window in time (in the 1970s), and brought on this interruption.

1. Planning Overhang. Reactor orders in the 1960s, based on a historic electrical demand growth rate of about 7%, far exceeded the actual power demand in the 1970s. The nation simply did not need the number of reactors in the construction backlog.

2. Lack of Standardization and Micro Safety Regulation. In the 1960s the nuclear reactor vendors were involved in a mega-watt race because station fixed costs for ancillary activities were essentially the same [and large] regardless of unit capacity. Also, technology was in a developmental stage, with changes being made constantly. The regulatory function was therefore as "ad hoc" as the designs being put before them for approval. The regulatory function understandably got involved in details of design, because the industry and the technology was not yet matured. It was a regime of "design and regulate as you go." This lead to unpredictable regulatory requirements, protracted construction schedules, and financial calamity.

3. Loss of Nuclear Power's Political Strength. Recall that the war in Vietnam had just ended, and the 1960s protest industry, idled at last, needed another target. Nuclear power was an attractive target: expensive, high tech, corporate-federal government ambiance, arrogant, a big program with the trappings of a closed club. In short order, in the mid 1970s, nuclear power lost its political infrastructure and aircover: demise of the the Atomic Energy Commission (US AEC), and the Congressional Joint Committee on Atomic Energy (JCAE), and then the election of an anti-nuclear president in 1976. President Carter attempted to halt breeder reactor and reprocessing development in the United States, and indeed in the rest of the world, and cast an agnostic spell on the then-operating nuclear power plants.

4. A Slack General Economy. The general economy was slack during this period, featuring high escalation rates and high interest rates, reaching double digits during the Carter Administration. This exacerbated the conditions already driving up nuclear capital costs, i.e. micro-regulation, and protracted construction schedules.

5. State Utility Commission (PUC) Pressures on the Rate Base. In the late 1970s and early 1980s the political pressures resulting from several nuclear project failures were intolerable. In many celebrated cases, PUCs disallowed excess costs in nuclear construction. Several plants promised at a few hundred million, were coming in at several billion, and the PUCs responded accordingly, financially ruining several utilities, and the careers of their leadership.

6. Loss of Confidence by the Financial Community. Seeing the effects of all of the above factors, the financial community withdrew. This was the final straw in the pyramid of woes facing commercial nuclear power in the 1970s.

7. TMI. The accident at TMI was a confirmation of the financial community's worst fears, and sealed the fate of the first wave of nuclear deployment in the US.

The above factors can all be classified as institutional/financial/political in character, having more to do with how a society organizes itself to accept, finance, develop, deploy, and regulate a new technology. The failure was not inherently technological.

While the US breeder program survived the Carter Administration, it did so just barely, and with a large expensive standing army marching in place. It was clear going into the Reagan Administration in 1981 that the severe fall-off in converter reactor deployment in the early 1970s eliminated the urgency for demonstrating and deploying the breeder and for moving ahead with reprocessing. An unusual coalition of fiscal conservatives and anti-nuclear activists in Congress brought down the Clinch River demonstration plant project in 1983.

ADAM SMITH AND THE ATOM

One of US nuclear power's principal impediments is that it is a creature of the US Federal Government. Nuclear power was born out of the Manhattan Program and the Naval Reactors Program in the 1940s and 1950s. In the early days, when the US Government was not an insular and gridlocked enterprise, things could actually get done--and get done accountably, professionally, and well. In the 1970s this changed. The Joint Committee was abolished and and its powers diluted into a myriad of Committee jurisdictions.

The Atomic Energy Commission vanished.

Continuity and quality of federal policy regarding nuclear energy suffered.

• The off-and-on federal policy regarding commercial reprocessing is an example. At least two of the three failed private sector attempts at reprocessing were frustrated by capricious federal policy.

• As noted above, the unpredictability of reactor licensing and regulation was one of the major factors leading to the nuclear recession which set in the early 1970s.

• In several cases, foolish restraints on exports of nuclear technology and equipment have hobbled the US vendors in their quest for commercial survival, through exports, in the 1970s.

•And, where the federal government has held statutory operational responsibility, it has generally failed catastrophically. Failure to execute the 1982 Waste Policy Act is an example. To make matters even more egregious, the current Administration actually uses the user-funded waste account as an optical deficit reduction device. The US uranium enrichment enterprise is another example of government fumbling, but happily this enterprise is on its way toward privatization.

The nuclear industry seems locked in a fatal embrace with its "nanny"---the federal government. It looks like, and in many cases *is*, a sick and dependent industry, unable to get along without federal largesse. Rather than fulfill its minimum statutory responsibilities expeditiously in the late 1980s ---efficient regulation, national waste system, and uranium enrichment---the government launched still another reactor R&D program on small "passively safe" reactors, at complete odds with the market place. When problems get intractable or merely difficult in executing its minimal statutory responsibilities, we find the government changing the subject to one more pleasurable, and spending more money, and creating more "pork-barrel" mortgages. This tilts the market place and unlevels the playing field; vendors whose products are not making it in the market place are made whole by the government.

What is needed from the federal government is stable, reliable, constructive, less intrusive policy--- *not money.* And the government must deliver on what it signed up to perform---deployment of a national waste system by a date certain. The US nuclear power infrastructure, in other words, needs a tincture, a drop, of Adam Smith restraint, accountability, and market place decision making.

I will bite my tongue and refrain from commenting on the current validity of the US Department of Energy.

THE NUCLEAR INTERREGNUM -- THE TURNAROUND GENERATION 1980 - 1995

Now permit me a personal note. The nation's current senior nuclear managers, now fifty-something who did not enjoy the development romance of the 1950s or commercialization boom days of the 1960s picked up the pieces in the 1980s and 1990s and sustained the nuclear option. We were not at the party, but we got the hangover. When we entered the nuclear power community working force in the early 1970s the bloom was off the rose, nuclear power's political and public popularity was vanishing, and US nuclear power was entering a long period of recession, rationalization, and maturation. This is not meant to start a generational war within nuclear power aficionados. It is simply a statement of fact. The forces causing a pause in nuclear power development and deployment were beyond anyone's control within the nuclear power community. My point is that we inherited a mess which was not of technological origins, and we ran with it,--we created a new policy based on market and political realities. This generation flogged through the realities of balance sheets, P&L statements, the immediate realities of cash flow. We consolidated, downsized & restructured, realigned product lines. We have forced legislative change in Washington. In other words we have made nuclear power a commercial reality, finally, as opposed to a technological dream , a haven for the endless stream of merchants of costly, undisciplined, un-market-driven R&D. And we are laying the pipe for a new nuclear era.

The "interregnum generation" has done fairly well:

- We have downsized, and restructured for real, not mythical, markets.
- Numerous consolidations have occurred in the industry to cope with the slack market---Framatome/B&W, Siemens/ANF, Framatom/Siemens, ABB/Combustion Engineering, etc. The nuclear supply infrastructure, anywhere you look, is at least twice the size of the served market. There are more consolidations to come.
- In my case (ABB Combustion Engineering), we have concentrated on the Asian market, and now have the dominant presence in South Korea with a backlog of six reactors, two of which are coming on line this year. General Electric has two reactors going up in Japan. Those of us who have regarded nuclear power as commerce in the national interest, rather than a technological sandbox, have done quite well.
- We have formulated and pushed through Congress regulatory reform legislation which will allow for preliscensing of standardized designs. During the Reagan Administration, in 1983, we submitted regulatory reform legislation to Congress, and this legislation was passed into law in 1992. The major US vendors have improved and standardized their designs. Combustion Engineering has been awarded an FDA for our System 80+ large standardized PWR design and GE has a similar achievement for its BWR design.
- Performance of operating nuclear plants in the US has been improved dramatically. Plant availability factors went up over the 1980s and 1990s, and the regulatory-driven excursion of operating & maintenance costs has been curtailed. The nuclear utilities have risen to a severe challenge, bitten the bullets, gotten their houses in order, and have succeeded.
- US nuclear utilities have taken charge of their own destiny, have insisted on and gotten more rational treatment from the US Nuclear Regulatory Commission.
- Nuclear waste legislation was passed in 1982, modified several times subsequently, and an actual, physical program to construct a geological repository has begun at Yucca Mountain, Nevada. This is a promise and responsibility not yet kept by a dilatory federal government.
- After facing down and turning around an acute market and commercial crisis in the early 1980s, the US uranium enrichment function is at last being privatized so that it can be operated like a business, rather than an instrument of foreign policy and a slush fund.

I could cite many more achievements of the "Interregnum Generation". In general it was a period of bullet-biting and problem solving, a period of facing market realities squarely, of working turnarounds, a period of transitioning US nuclear power from a hobby-shop to a viable business sector and, therefore, a sustainable energy option for the future.

THE NEXT FIFTY YEARS

Now, let's turn to the crystal ball.

Macroeconomics and geopolitical imperatives will force a second wave of nuclear deployment here and abroad. Duke Power CEO Bill Lee gave a spellbinder at the 1992 ceremony marking the 50th anniversary of the Stagg Field experiment. His speech identified several key factors which auger for a second coming:

- environmental constraints on fossil fuel burning, most notably, the greenhouse effect and global warming, and implementation of the Clean Air Act.
- depletion of fossil fuels;·
- population growth and demand for even higher standards of living in the developed world;
- in the undeveloped world, population growth and demand for standards of living on par with that currently available in the developed world;
- the inability of conservation alone, to fill the gap;
- and the absence of alternatives to nuclear (fission) power visible now.

Without a major role for nuclear power over the next 50 years, the numbers simply do not add up, and this imperative will somehow find its way into the body politic, the public consciousness and into the market-place, if calamity is to be avoided.

My belief is that Asia will continue to be the largest market for nuclear power over the next several decades, and will be a safe haven for US-origin technology and know-how. Western Europe, especially France, will continue to be a stalwart, abiet with some falloff in nuclear power capacity growth rate.

In the US, a re-ignition of nuclear capacity growth will be much more sober than the first nuclear wave (1960s). The US enjoys a market economy, and large, capital-intensive, long-term enterprises like nuclear power are inertial and have trouble adapting swiftly in such economies. There are several prerequisites which are being put in place now, such as prelisensing of standardized designs in a more streamlined and effective regulatory regime.

The *largest challenge for a new US deployment will be in reshaping the commercial structure,* i.e. who will own and operate the next generation of nuclear plants in the US? It is clear to me that in the next generation, vendors will have to assume more risk than they have assumed in the current mode where the utilities (through their AE surrogates) assume the responsibility of overall design ,construction, and project management.

There are several conceptual models to consider:

1. <u>The Nuclear Monopoly</u>. This is the French model. This model inherently involves federal ownership and management, or, at least, enabling sanctions and federal backstopping. The nation would have one technology, and one institution per function in the nuclear energy system--one nuclear utility, one reactor supplier, one fuel cycle company, and so on.

This arrangement has a certain theoretical and emotional appeal--it suggests a national call to arms to preserve and protect the nuclear option, and indeed it answers a substantive requirement for economies of scale and financial viability. The "nuclear monopoly" model has worked quite well in France, but then France has an entirely different set of energy imperatives than the US. France and Japan do not have the luxury of numerous energy options and abundant fuel resources. Although we occasionally hear calls for a nuclear monopoly approach in Washington, it is basically a cop-out, an admission that nuclear power cannot stand the Adam Smith crucible of the marketplace.

2. Status Quo. In this model, the utility company assumes all construction, operational, licensing, schedular, financial risks. This was the traditional format that the US power industry settled back into after the ruinous early rounds of vendor turnkey projects.

Even in this model, the vendor will assume more risk than he has in the past: in the new NRC licensing format, vendors will own the certified designs, not the utilities, and the vendors will be accountable for complete designs.

3. Turnkey. This was the model used when the first few nuclear plants were ordered in the early 1960s. Vendors took complete responsibility for designing, licensing, and construction, and then "turned over the keys" to the utility when the plant was completed. Cost and schedular risks were assumed by the vendor. This scheme proved ruinous to several US vendors because, in those days nuclear power was in a developmental phase, capital cost pressures drove the industry into a "mega-watt race", and the regulatory standards were a moving target.

However, the turnkey model would involve less vendor risk now than in the early days. The industry has matured technologically. The "mega-watt" race is long over. The provisions in the 1992 Energy Policy Act allow for prelisensing of standardized designs, meaning that schedular and other financial risks can be abated.

4. Vendor Equity Positions. A step significantly beyond turnkey is the prospect of vendors taking equity positions in a nuclear plant--giving them a share in ownership during construction and possibly during operation. Return (or loss) on this equity would be realized by ultimate sale of shares or by participation in the revenue stream from electricity sales.

This model exposes the vendor to operational performance risks, as well as construction costs and schedule risks. It keeps the vendor glued to the project longer, although it blurs traditional vendor vs utility roles, and dilutes "arms length" relationships.

5. Nuclear Independent Power Producers (NIPPs). The NIPP involves the most dramatic departure from traditional relationships. The IPPs would be established as wholesale generators of electricity, and, operating interstate, would theoretically escape capricious state rate regulation. The generators, perhaps owned by reactor suppliers, utility subsidiaries, architect-engineers, would sell electricity to the utilities whose role would be reduced to transmission and distribution.

The design of a new institutional/financial structure for nuclear power stewardship is, in my mind, the central challenge,.rather than development of a new reactor type. Technology is not the barrier---institutional, financial, political factors are. This and other institutional issues are what we should put our shoulders and minds to and to be creative about---not the expensive and irrelevant invention of new and better technological paper mouse-traps.

The next US wave of orders I would expect to start in the next 15 years, and I would expect that orders will gravitate to large, standardized, evolutionary, pre-licensed LWR designs, rather than a new reactor species. For one thing, federal largess in bankrolling a new reactor type will not occur in the next several decades. We saw the last [for the foreseeable future] of the big demonstration projects with the cancellation of Clinch River in 1983.

What then of the breeder? How will we tap into the U 238 natural endowment (representing 99.3% of the uranium energy content)? Fortunately, the U 238 inventory does not perish, and it can be tapped at our will and at our necessity. The asset remains in the depleted tails from enrichment operations, stored in canisters at Oak Ridge, Paducah, and Portsmouth, and it does not matter whether we mine this asset, as well as the uranium left in the ground, now or later. I am assuming of course that Uncle Sam does not do something stupid like destroy this asset; given Washington's past performance, there is risk in making this assumption.

So it boils down to the timing of the need for a breeder, and this has been the central issue from day one 30 years ago. Clearly, the assumptions that lead to the expensive and aggressive demonstration program formulated in the 1960s were wrong---1200 GWe of nuclear installed capacity by the year 2000. The LWR order recession that set in in 1974 invalidated that assumption. But you quickly confront a paradox in matters like this: it's like asking a nine-month pregnant woman to hold it in for just another few years. We had an expensive standing army marching in place when President Reagan was elected and ordered remobilization of the breeder program, following the failed Carter presidency.

The timing of the need for a breeder is further complicated by the weapons fissile material overhang. The various disarmament treaties of the late 1980s and early 1990s have resulted in excess inventories of Highly Enriched Uranium (HEU) and Plutonium (Pu). This overhang threatens the viability of uranium mining and enrichment enterprises, and adds at least 15 years to the date when we must get serious about a breeder economy.

Given the imperatives that Bill Lee outlined in his 1992 speech, I would place the need for a breeder economy sometime in the second quarter of the next century. By then we have may have lost the race memory of breeder technology. When the final defeat for Clinch River (a deeply impaired project) came on the Senate floor in 1983 , I hurried to put in place an agreement with Japan (MONJU) for a sustaining US access to the technology. While I tried to make a deal with Japan or South Korea on the billion plus US investment in CRBR components, we were directed to destroy these components and leave a scorched earth, as if the national breeder program never existed. When we need essential breeder technology in the future, we can get it from Japan and France, but with some loss in energy independence.

SUMMARY

In summary, I expect a "second coming" of nuclear power during the next half century, because of the driving geopolitical and environmental imperatives cited by Bill Lee in his 1992 speech. The factors resulting in the truncation of the first era, suggest that the challenges we will face in the next nuclear era will be institutional, political, and financial, rather than technological.

THE WORLDWIDE PERSPECTIVES FOR NUCLEAR POWER

Poong Eil Juhn

Director, Division of Nuclear Power
Department of Nuclear Energy
International Atomic Energy Agency
P.O. Box 100
A-1400 Vienna, Austria

ABSTRACT

Nuclear energy is a proven technology that already makes a large contribution to electricity supply worldwide. At the end of 1994, there were 432 nuclear power plants (NPP) operating in 30 countries with a total capacity of some 340 GW(e). About 2,130 TW.h were produced from these NPPs, which shared over 17% of total world electricity production in 1994. The paper outlines the global perspectives for nuclear power worldwide, drawing upon the information collected by the International Atomic Energy Agency (IAEA) and integrated in its data bases.

The paper also analyses the trend in electricity consumption growth of the world. This analysis indicates that economic development and improved quality of life are closely linked with the availability of energy and electricity services. Changes in lifestyle and consumption patterns are likely to induce more efficient use of energy since sustainability has become a major objective. However, population growth, urbanization and industrial development will increase the demand for electricity especially in developing countries and call for development and deployment of all the technologies available. In this connection, the paper analyses the opportunities and challenges for a broader deployment of nuclear power worldwide.

Although the development of nuclear power has slowed down in most countries during the recent years, 48 nuclear power plants under construction in 15 countries, with a total capacity of 39 GW(e), remain significant. Taking into account uncertainties related to construction and licensing lead times, the worldwide increment of nuclear capacity by the turn of the century may range from 8% to 10%, leading to a total installed capacity of some 367 to 375 GW(e).

In the medium term, up to 2015, the national energy plans give a reasonable basis for estimating nuclear power development, even though uncertainties regarding electricity demand forecasts, policy decisions, availability of funding, etc. are involved. The low and high cases developed by the IAEA, in which nuclear share in electricity generation in 2015 is assumed to be 12% and 14%, respectively, show that nuclear capacity in 2015 would be in the range from 369 GW(e) to 516 GW(e), which will produce from about 2400 to 3300 TW.h of electricity.

Economics and Politics of Energy
Edited by Kursunoglu *et al.*, Plenum Press, New York, 1996

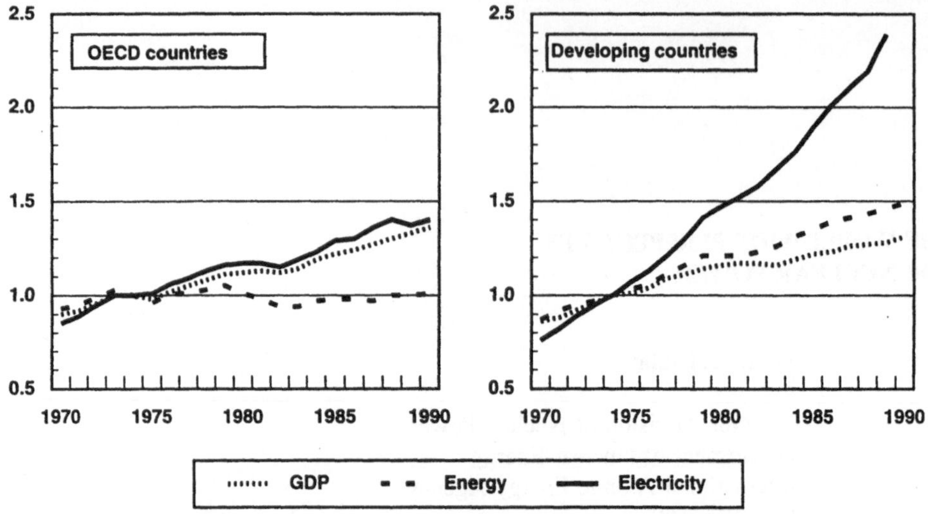

Figure 1. Growth trends for Gross Domestic Product (GDP), energy and electricity. (Per capita values; normalized to 1.0 in 1974)

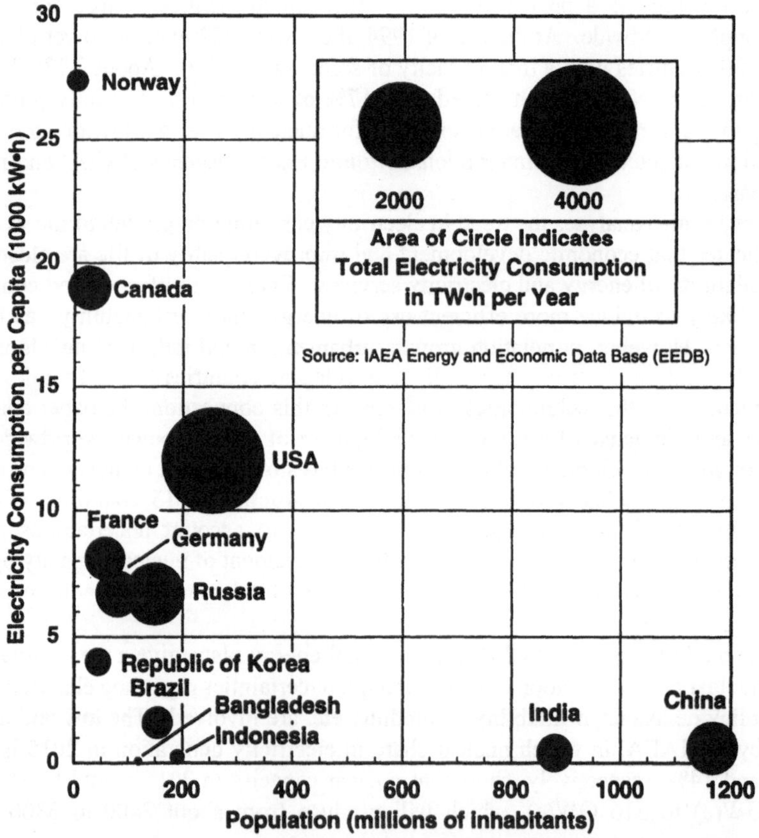

Figure 2. Annual electricity consumption versus population, for selected countries. (1992 data)

PROSPECTS OF ENERGY AND ELECTRICITY

Energy is a fundamental part of everyday life. Electricity is in these days the most convenient and versatile form of energy. Increased use of electricity is an important factor in modernization and in achieving greater efficiency in total energy use. Techno-economic studies have clearly shown a high correlation between electricity consumption and national economic output for a wide range of countries, with the obvious conclusion that developing countries will need a large expansion of their electricity-producing capacity if their aspirations for economic growth and higher standards of living are to be achieved. Full participation in the information and communication age, which the world is now entering, will require the availability of reliable sources of electricity. Even when total energy consumption is not rising, or in countries where it is even declining, data shows that electricity consumption continues to grow (Figure 1). From 1960 to 1990 the share of electricity has grown from 17 to 30% even considering the fact that two billion people in the world still do not have access to electricity in their homes. The worldwide phenomena of urbanization, allowing easier access to electrical distribution systems, together with the electrification of rural areas, will result in a still greater share for electricity in the future. The world per capita annual consumption of electricity has risen about 3 times from 1960 to 1990 (from 765 to 2225 kWh per person)[1].

If we look at per capita electricity consumption in individual countries, as illustrated in Figure 2, highly populated developing countries at present consume much less electricity per capita than do developed countries. It is quite obvious that per capita consumption of electricity in developing countries has to increase substantially, if increased economic growth and improved standard-of-living are to be achieved. The example of the Republic of Korea, where per capita electricity consumption has grown from 70 kWh per year in 1960 to almost 3200 kWh per year in 1992, clearly shows that large changes in economic prosperity, with correspondingly higher demands for electricity are possible. The future electricity growth will likely be dominated by such developing countries as Brazil, China, India, Indonesia and Pakistan, in their achievement of economic growth with large populations.

For the next decades, it is expected that the developing countries, which will account for 95% or more of the world's population growth (Figure 3), will endeavour to enhance their economic and social development. This will increase drastically the demand for electricity from the industrial, service and residential sectors, requiring substantial additional power production capacities, even taking into account demand management policies and productivity gains.

The primary means of producing electricity and their contributions to world electricity production are shown in Figure 4. As seen, fossil power, that is the burning of coal, oil and gas, is the dominant means of producing electricity in the world today, contributing almost 63% of the total. Hydro energy follows with 19.5% and nuclear with 17%. Contributions by other means of geothermal, solar, wind and bio-mass are not now significant, less than 0.5% all together[2].

The decisive factors which influence decisions regarding the construction of electricity generating facilities vary widely from region to region and with time. Decision factors which traditionally have been important are local availability of fuel, generation reliability and overall cost. Other factors include: security of supply, planning flexibility, environmental impact and capital requirements. One of the major impacts on the energy planning process in many countries is to reduce the need for new generating facilities by promoting energy conservation, demand management and efficiency improvement. However, the construction

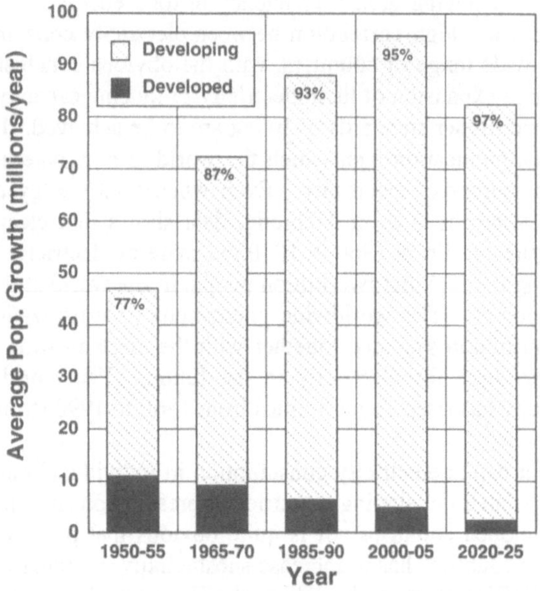

Figure 3. Population growth in developed and developing regions of the world.

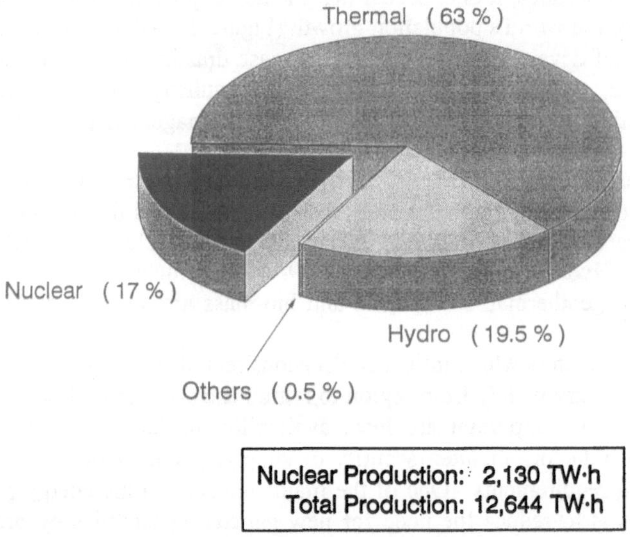

Figure 4. Fuel shares in 1994 world electricity production.

of new facilities is inevitable, to meet, as discussed earlier, the needs of increasing populations and enhanced economic growth, as well as for replacement of current facilities at their end of life. Worldwide consumption of electricity can be expected to increase by 50% to 75% by the year 2020, according to the World Energy Council[3].

CURRENT STATUS OF NUCLEAR POWER DEVELOPMENT

Figure 5 shows the growth of total and nuclear electricity generation since 1970. Nuclear electricity generation expanded rapidly in the 1970's and by 1980 had reached 692 TW.h, over nine-fold increase since 1970, and contributed 8.4% of the total electricity production. In the period 1980-1985 nuclear electricity generation increased in two times to some 1400 TW.h, which shared 14.3% of the total electricity. During 1985-1990, nuclear generation increased to 1913 TW.h which shared 16.2% of the total electricity production. In the period 1990-94, nuclear generation has been slowed down to 2130 TW.h, which shares 17% of total electricity supply.

According to data in the IAEA's Power Reactor Information System (PRIS)[4], at the end of 1994 there were 432 NPPs connected to electricity supply networks in 30 countries (Figures 6 and 7), with a total installed nuclear power generating capacity of 340.4 GW(e). There were also 48 NPPs under construction in 15 countries (Figures 8 and 9), with a total generating capacity of 39 GW(e). Accumulated nuclear reactor operating experience reached approximately 7230 reactor years. Worldwide, in 1994, 32 countries (including Taiwan, China) were operating or building a total of 480 NPPs for electricity generation.

During 1994, four NPPs, having a total net capacity of 3356 MW(e), were connected to the grid in four countries: China (1), Japan (1), Republic of Korea (1) and Mexico (1), increasing the world's total installed nuclear power generating capacity by about 1%.

Thus, nuclear power is a proven technology which already makes a large contribution to electricity supply, amounting to about 17% of the total electricity generation in the world and, to a much lesser extent, to heat supply in some countries. Moreover, nuclear power is economically competitive with fossil fuels for base load generation in many countries, and is one of the commercially proven options that already is making a substantial contribution, which could be enlarged in the future, to reduce environmental burdens, especially greenhouse gas emissions, from the electricity sector. Worldwide, eighteen countries (including Taiwan, China) relied upon nuclear power plants to supply over 25% of their total electricity needs (Figure 10). Lithuania and France have the largest shares of nuclear power, with more than 75% of their total production, followed by Belgium with 56%. The largest nuclear electricity producers in the world are the USA (639.4 TW.h), France (342 TW.h), Germany (143 TW.h), Japan (258 TW.h), Canada (102 TW.h) and Russia (98 TW.h).

Since global climate change is considered to be a major environmental threat, greenhouse gas emissions from electricity generation chains are the focus of attention of energy planners and decision makers at the national and at the utility level. Recent studies on greenhouse gas emissions from different energy chains for electricity generation indicate that nuclear power is one of the better options to alleviate global climate change (Figure 11). Renewable sources emit in the range of 2 to 5 times more greenhouse gases than nuclear power per unit of electricity produced when the entire fuel chain is considered, and the emissions from fossil fuel chains range from 40 to 100 times more!

The role that nuclear power can, and indeed already does play, in reducing global emissions of greenhouse gases, in particular CO_2, is illustrated in Figure 12 showing that nuclear power is today avoiding some 8% (about the same as hydropower) of additional CO_2

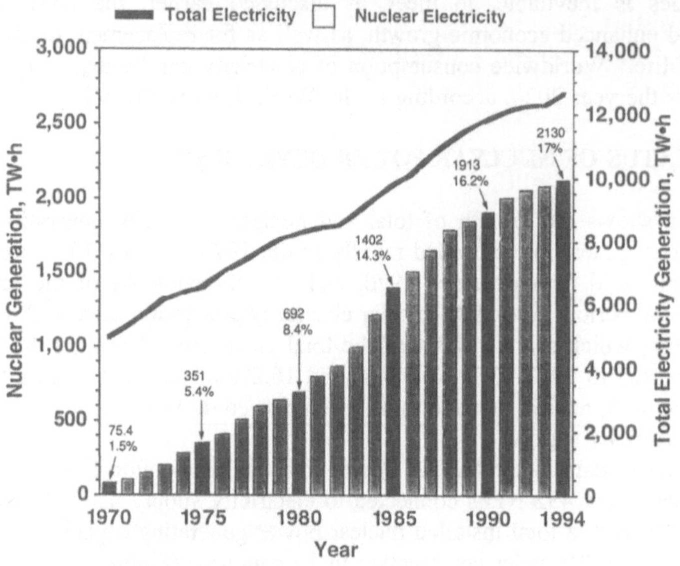

Figure 5. Growth of nuclear generation since 1970.

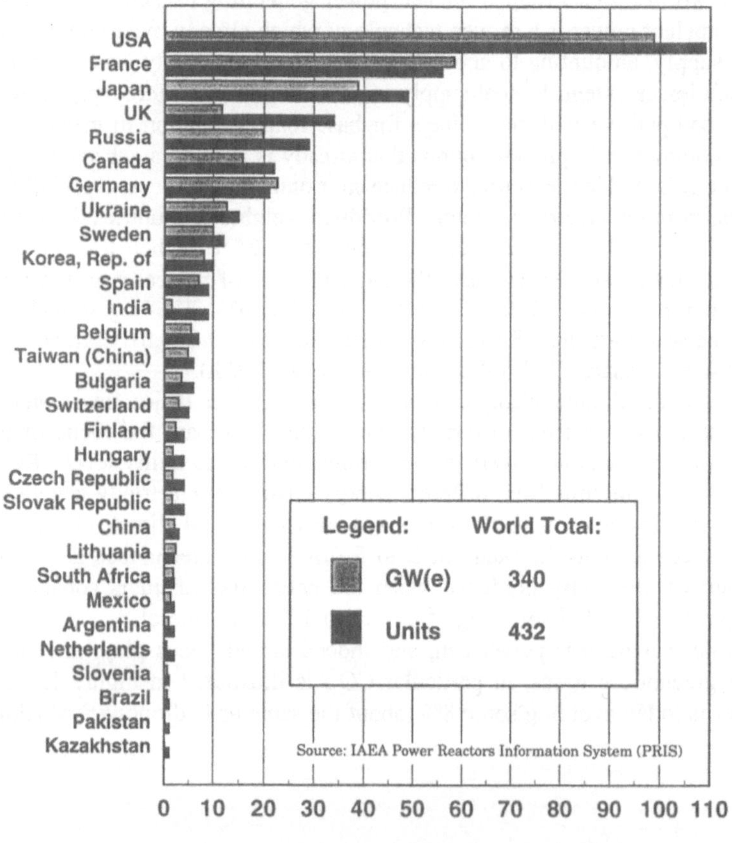

Figure 6. Nuclear power plants in operation at end of 1994.

emissions that would have occurred if the electricity produced by nuclear power were to be produced by fossil fuels.

Nuclear power alone will not ensure secure and sustainable electricity supply worldwide, nor will it be the only means of reducing greenhouse gas emissions, but it has a key role to play in this regard. The challenge for realizing the necessary revival of the nuclear option is to improve the technical and economic performance of nuclear power plants while enhancing even further their safety and satisfactorily addressing the issues of waste management and disposal.

Natural resources, technologies and industrial capabilities are available to sustain a large scale deployment of nuclear power. Uranium supply is not likely to be an issue in the coming decades even if a major revival of the nuclear option would occur. In the short and medium term, existing uranium production capacities and civil and military stockpiles will be more than sufficient to ensure security of supply at the front end of the fuel cycle for the nuclear power plants already operating and expected to be built in the coming decades. Improved fuel design and management can reduce the specific uranium requirement per unit of nuclear electricity generated.

The advantage of nuclear power can be further increased by the efforts of reactor designers to reduce capital costs by streamlining the reactor concepts, reducing the amount of material required, and shortening the construction times. Substantial progress has been achieved in this regard and additional gains are expected through the development of advanced reactors. However, financing the large capital costs of nuclear power plants will remain a key issue in developing countries. Technology adaptation, the development of small and medium size reactors, and the implementation of new financing approaches may alleviate the funding constraints and facilitate a broader deployment of nuclear power in developing countries.

FUTURE NUCLEAR POWER PROJECTIONS AND VIEWS

Projecting future nuclear power development is a somewhat difficult exercise, since a number of factors which may influence policies, decision making and implementation of programmes cannot be assessed with certainty.

Up to the year 2000, the installed nuclear capacity worldwide will grow to between 367 GW(e) and 375 GW(e), compared to 340 GW(e) in 1994[5]. Since all the units to be commissioned by the turn of the century are already under construction, the range of uncertainty reflects potential delays in construction and licensing. New nuclear units will be connected to the grid mainly in Asia, while in Western Europe and North America the installed nuclear capacity will remain practically unchanged. In Eastern Europe, although some of the units under construction will be completed, the economic transition will delay significantly the implementation of the nuclear programmes in most countries.

After the turn of the century, the range of uncertainty regarding nuclear power development is wider owing to a number of technical, economic, environmental and policy factors. The low and high nuclear generating capacity projections developed by the IAEA up to 2015 are based upon a review of nuclear power projects and programmes in Member States (Figure 13). They reflect contrasting but not extreme underlying assumptions on the different driving factors that have an impact on nuclear power deployment. These factors and the ways they might evolve vary from country to country. The IAEA projections are not predictive and do not reflect the whole range of possible futures from the lowest to the highest feasible, but provide a plausible range of nuclear capacity growth by region and worldwide.

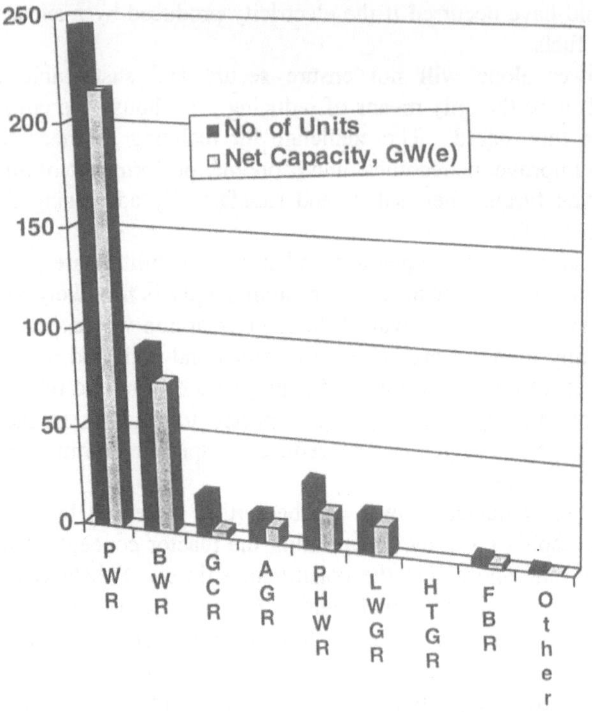

Figure 7. Nuclear power plants in operation at end of 1994.

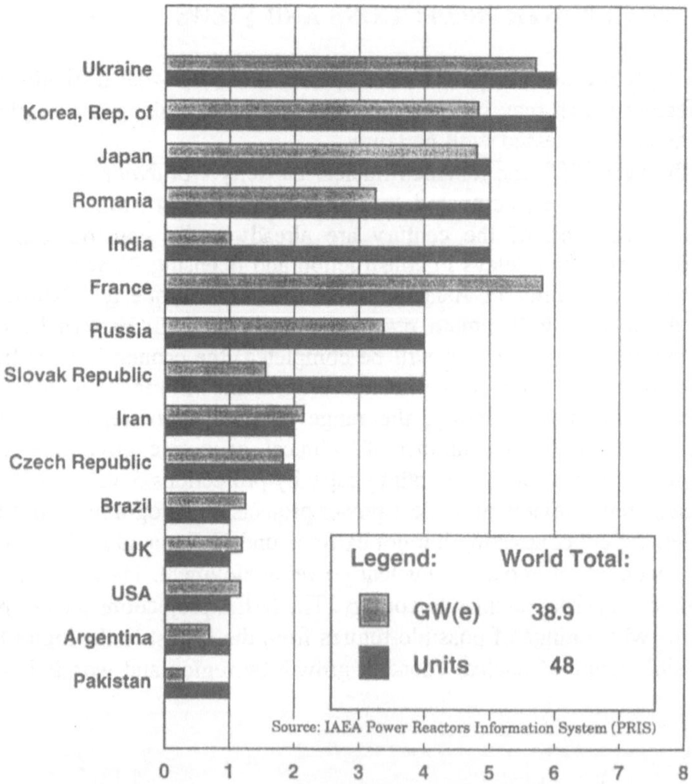

Figure 8. Nuclear power plants under construction at end of 1994.

In the low case, the present barriers to nuclear power deployment are assumed to prevail in most countries during the coming two decades. Economic and electricity demand growth rates remain low in industrialized countries. Public opposition to nuclear power continues, and environmental concerns such as the risk of global climate change do not become strong driving factors in energy policies aiming towards switching from fossil to nuclear energy. Institutional and financing issues prevent the implementation of previously planned nuclear programmes in particular in countries in transition and in developing countries. There is no drastic enhancement regarding nuclear technology adaptation and transfer, nor financial support to developing countries for the implementation of nuclear power projects. Under these rather pessimistic assumptions, most of the nuclear units under construction would be completed but new nuclear units would be ordered only in the countries where nuclear power is a major component of electricity generation mixes, such as France, Japan and the Republic of Korea. Owing to the large number of units that would be shut down at the end of their scheduled operating lifetime, the total nuclear capacity in the world would start to decrease after 2010 and would be similar in 2015 to that in 2000, i.e., some 370 GW(e). The share of nuclear power in the world electricity supply would decrease from about 17% at present to some 13% in 2015.

The high case reflects a moderate revival of nuclear power development, that could result in particular from a more comprehensive comparative assessment of the different options for electricity generation, integrating economic, social, health and environmental aspects. This case assumes that some policy measures would be taken to facilitate the implementation of these programmes, such as strengthening of international co-operation, enhanced technology adaptation and transfer, and establishment of innovative funding mechanisms. With these assumptions, the total installed nuclear capacity worldwide would reach some 515 GW(e) in 2015 and the share of nuclear power in total electricity generation would be some 15%.

In both the low and high cases, the production capabilities of the world nuclear industry would exceed the demand for new reactors. A higher rate of nuclear power development would be technically feasible and economically viable in a number of countries. However, a substantive revival of nuclear power programmes would require policy measures, including a removal of the de facto moratoria in several countries and the introduction of mechanisms for providing funding support to nuclear projects in developing countries, which seem unlikely to be implemented in the short term.

NON-ELECTRICAL APPLICATION OF NUCLEAR ENERGY

Currently only a few nuclear plants are being used for non-electric applications (with a total capacity of only 5 GW(th) to supply hot water and steam). However, at present, about 30% of the world's primary energy consumption is used for electricity generation, about 15% is used for transportation and the remaining 55% is converted into hot water, steam and heat. This shows that the potential market for applications of nuclear energy in the non-electric energy sector may be quite large. Non-electric applications include desalination, hot water for district heating, heat energy for petroleum refining, for the petrochemical industry and for the conversion of hard coal or lignite. As shown in Figure 14, for non-electric applications, the specific temperature requirements vary greatly. Hot water for district heating and heat for seawater desalination require temperatures in the 80 to 200°C range. Temperatures in the 250 to 550°C range are required for petroleum refining processes. The use of heat for enhancing heavy oil recovery can be applied by the method of hot water or steam injection.

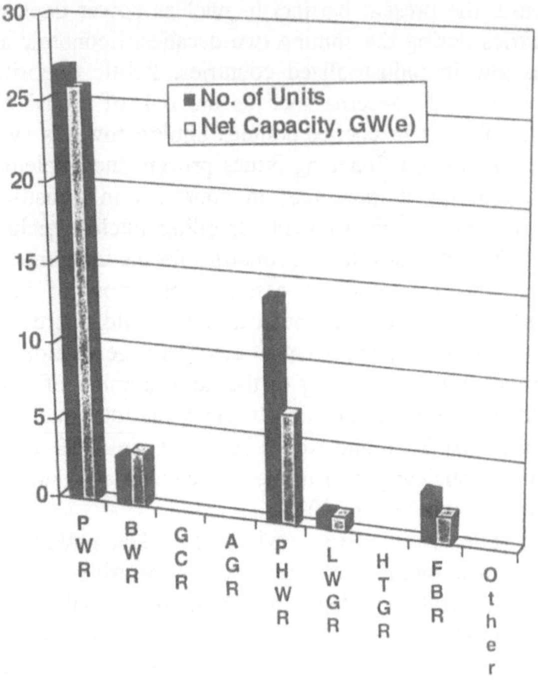

Figure 9. Nuclear power plants under construction at end of 1994.

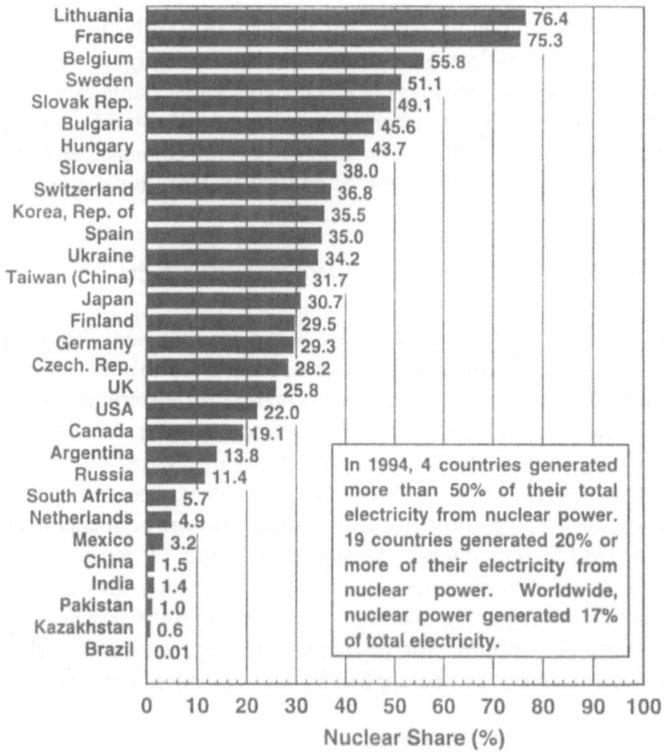

Figure 10. Nuclear Shares of Total Electricity Generation– 1994.

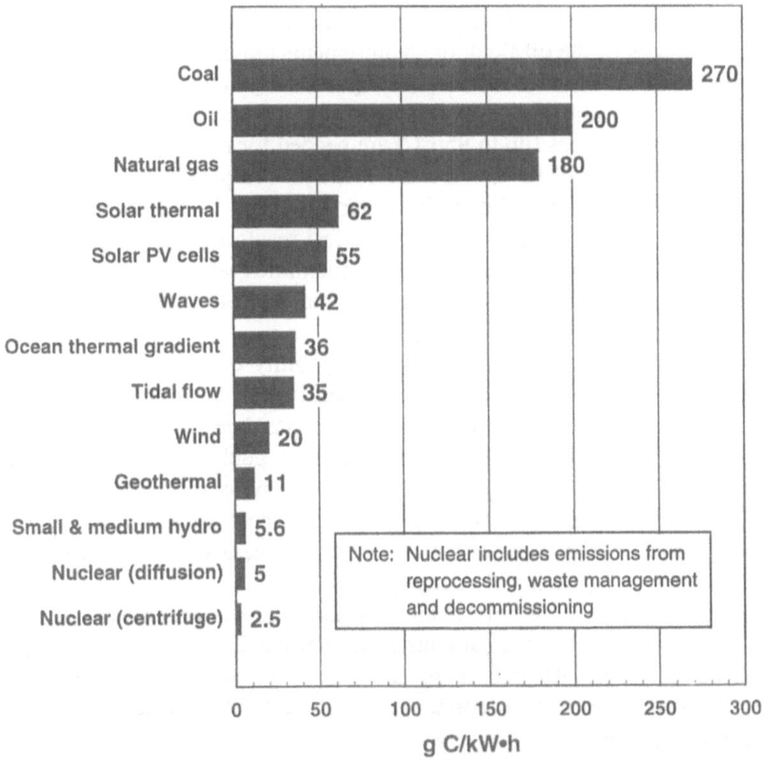

Figure 11. CO_2 emission factors from the full energy chains of different power sources. Source: "Atoms in Japan," November 1994.

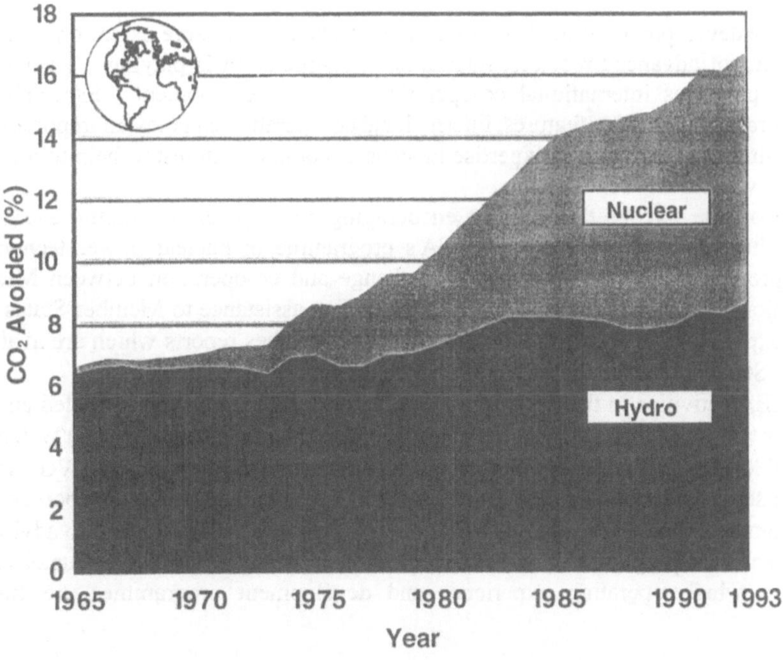

Figure 12. Percentage of CO_2 avoided globally by nuclear and hydro energy.

The temperature and pressure conditions required for heavy oil recovery are highly dependent on the geological conditions of the oil field, the requirements ranging up to 550°C and above. Temperatures required for oil shale and oil sand processing range from 300 to 600°C. Processes used in the petrochemical industry require higher temperatures, in the range of 600 to 880°C. Still higher temperatures (up to 950°C) are needed for refining hard coal or lignite (for example, to produce methanol for transportation fuel). Temperatures of 900 to 1000°C are necessary for the production of hydrogen by water splitting. Water-cooled reactors can provide heat up to about 300°C. Liquid-metal-cooled fast reactors produce heat up to about 540°C. Gas-cooled reactors provide even higher temperatures, about 650°C for advanced gas-cooled, graphite-moderated reactors (AGRs), and up to 950 to 1000°C for high-temperature gas-cooled reactors (HTGRs).

There is considerable incentive to utilize the capability of nuclear plants to provide co-generation of electricity, steam and heat for residential and industrial purposes[6]. Experience in co-generation with water-cooled reactors has been gained in the Russian Federation, China, Canada, the Czech and Slovak Federal Republic, Switzerland, Germany, Hungary and Bulgaria. One of the largest uses of nuclear process steam is at the Bruce Nuclear Power Development Facility in Ontario, Canada, where the CANDU PHWRs are capable of producing 6000 MW(e) of electricity as well as process steam and heat for use by Ontario Hydro and an adjacent industrial energy park.

An important milestone in the development of high-temperature nuclear process heat was reached in March 1991 with the start of construction of the 30 MW(th) high temperature test reactor (HTTR) of the Japan Atomic Energy Research Institute. The HTTR will be the first nuclear reactor in the world to be connected, at around 1998, to a high-temperature process heat utilization system.

ROLE OF THE IAEA IN NUCLEAR POWER

Advanced Reactor Development

The early development of nuclear power was conducted to a large extent on a national basis. However, for advanced reactors, international co-operation is playing a greater role, and the IAEA promotes international co-operation in their development. Especially for designs incorporating innovative features, international co-operation can play an important role allowing a pooling of resources and expertise in areas of common interest to help to meet the high costs of development.

To support the IAEA's functions of encouraging development of atomic energy for peaceful uses throughout the world, the IAEA's programme in nuclear power technology development promotes technical information exchange and co-operation between Member States with major reactor development programmes, offers assistance to Member States with an interest in exploratory or research programmes, and publishes reports which are available to all Member States interested in the current status of reactor development.

The IAEA activities in the development of water cooled, liquid metal cooled and gas cooled reactors are co-ordinated by three international working groups (IWGs) which are committees of leaders in national programmes in these technologies. Each IWG meets periodically to serve as a global forum for information exchange and progress reports on the national programmes, to identify areas of common interest for collaboration and to advise the IAEA on its technical programmes and activities. This regular review is conducted in an open forum in which operating experience and development programmes are frankly discussed.

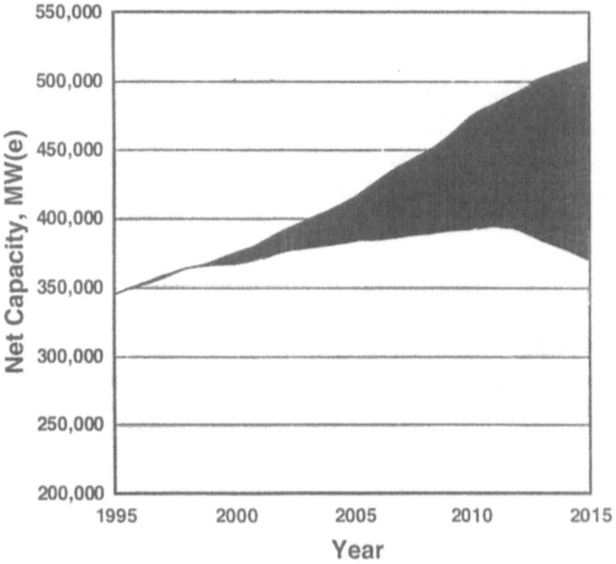

Figure 13. Worldwide nuclear power outlook up to 2015.

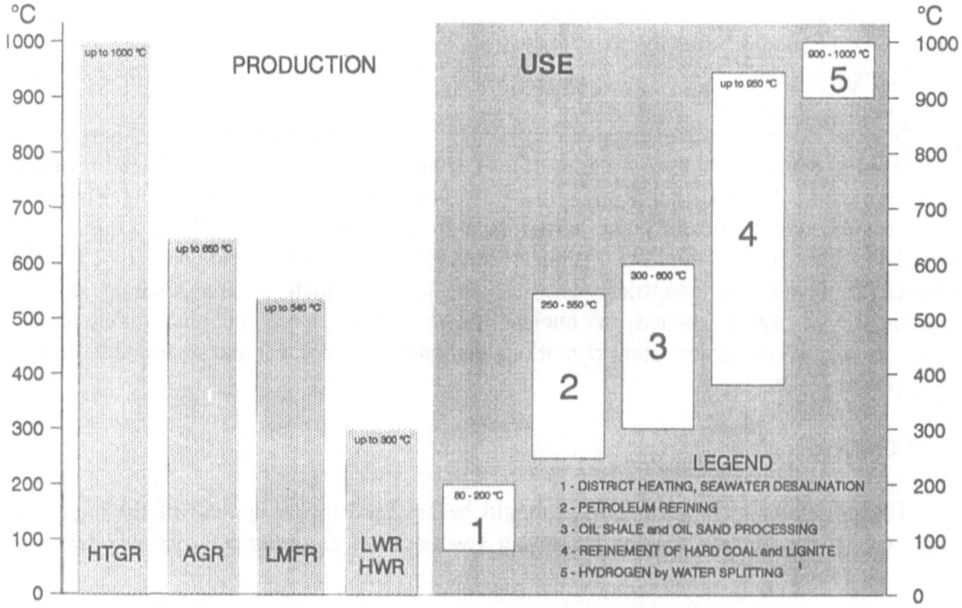

Figure 14. Temperature ranges in production and use of nuclear energy.

Small specialists meetings are convened to review progress on selected technology areas in which there is a mutual interest. For more general participation, larger technical committee meetings, symposia or workshops are held. The IWGs sometimes advise the IAEA to establish co-operative programmes in areas of common interest in order to pool efforts on an international basis. These co-operative efforts are carried out through co-ordinated research programmes (CRPs). CRPs are typically of a duration of 3 - 5 years and often involve experimental activities. Such CRPs allow a sharing of efforts on an international basis to develop technology at a lower cost than would be required with separate national efforts, and to benefit from the experience and expertise of researchers from the participating institutes.

Nuclear Power Plant Project Management

The IAEA provides assistance to requesting Member States on a variety of areas related to the use of nuclear power for safe and economic electricity generation. This assistance is directed to the national authorities, operating organizations and the industrial infrastructure that supports the planning and execution of nuclear power project.

An important objective is to assist developing Member States in achieving self-sufficiency in the systematic development and improvement of a wide range of infrastructural facilities, including organizational structures and their management, transfer of tools and methodologies for project management and the feedback of construction experience from successful project execution.

Not less important is the problem of providing qualified human resources to plan, regulate and implement the nuclear power programme. Thus manpower development programmes are being supported whit the objective of creating an independent national training infrastructure that meets the needs of the nuclear industry. To this effect ten projects for nuclear power programme implementation are being serviced in eight countries in the Middle East-Europe and Asia and Pacific regions.

In order to ensure that the functional capability of critical plant equipment remains within the design intent, periodic in-service inspections, testing programmes and surveillance functions are necessary. These activities are important as they provide a guidance for good plant life management and direct executive decisions for technical and economic assessments of the viability of plant life extension.

In order to complement assistance provided under individual projects, a number of topical training courses, workshops and seminars ar held on subject matters selected for their relevance to country specific needs and requirements. At the international level major training courses are offered periodically for participants from all the regions. These courses are designed to represent and effective transfer of technology that helps create a base of trained personnel in developing countries. They cover subjects such a strengthening project management, quality assurance in nuclear power plant operation and maintenance, qualification of nuclear power plant operations personnel and control and instrumentation in nuclear power plants.

CONCLUSIONS

The prospects for nuclear power should be assessed in the context of the expected growing electricity demand and the increasing awareness of environmental issues. Nuclear power alone will not solve all the problems, and ensuring sustainable electricity supply worldwide will require the development of different energy systems adapted to the resource base and economic situation of various regions. However, nuclear power does have the potential to contribute significantly to optimized electricity system expansion strategies based upon economic and environmental criteria.

Whether nuclear power is the cheapest option will differ from country to country and will be strongly influenced by the price of fossil fuels and by the capital costs. In order for the nuclear power to be competitive, it is essential that nuclear projects be well managed during construction and operation. In order to further improve economics of nuclear power, new designs of nuclear power plants should lead to plants that are simpler, easier and cheaper to construct, which can be achieved by the application of standardization and modularization of plants.

It is expected in the near future that nuclear power projects will continuously be promoted mostly in Asian countries including China, Japan and the Republic of Korea. However, in many other regions of the world, safe and reliable operation of nuclear power plants, convincing solutions of high level nuclear waste storage and disposal, and a predictable licensing process are essential prerequisites for the revival and expansion of nuclear power, which is viable, technically proven, economically competitive and environmentally benign.

REFERENCES

1. P.E. Juhn, *Global Perspectives for Nuclear Power,* 10th Biennial Nuclear Conference, organized by ENFIR, Águas de Lindóia, Brazil (1995).

2. International Atomic Energy Agency (IAEA), *Energy, Electricity and Nuclear Power Estimates for the Period up to 2015,* Reference Data Series No. 1, Edition (1995).

3. World Energy Council, *Energy for Tomorrow's World: the Realities, the Real Options and the Agenda for Achievement,* St. Martin's Press (1993).

4. International Atomic Energy Agency (IAEA), *Nuclear Power Reactors in the World,* Reference Data Series No. 2, Edition (1995).

5. International Atomic Energy Agency (IAEA), *IAEA Yearbook,* IAEA (1995).

6. International Atomic Energy Agency (IAEA), *Nuclear Applications for Steam and Hot Water Supply,* IAEA-TECDOC-615 (1991).

A DOE PERSPECTIVE ON THE FUTURE OF NUCLEAR ENERGY

Terry R. Lash, Director,
and William D. Magwood, IV,
Associate Director for Planning and Analysis

Office of Nuclear Energy, Science and Technology
U.S. Department of Energy
Washington, D.C. 20585

INTRODUCTION

This is an important stage in the history of nuclear energy. The Tennessee Valley Authority's (TVA) Watts Bar 1 nuclear power plant, apparently the last U.S. nuclear power plant of the current generation, was granted a full-power operating license by the Nuclear Regulatory Commission (NRC) in February 1996, and is expected to begin commercial operation in early 1996. At the same time, the first two designs of the next-generation plants have both received their Final Design Approvals from the NRC and are expected to receive design certifications before the end of 1996.

In the rest of the world, the timing is no less interesting. Nuclear programs in European countries are slowing even as the issue of how to reduce carbon dioxide emissions moves to the forefront. In Asia, the world's center of nuclear construction activity for most of the last decade, the drive toward building new nuclear plants has accelerated, due mainly to the announced desire of the People's Republic of China to build 50 gigawatts of new nuclear generation capacity over the next 25 years. Finally, in the nations of Eastern Europe and the Former Soviet Union, the long, difficult process of enhancing the safety of Soviet-era reactors and building the infrastructures and cultures necessary to increase and sustain adequate levels of nuclear safety has begun in earnest.

Whatever the future of nuclear proves to be during the coming century, the foundations are being laid now. We believe that there is a clear opportunity for new nuclear plants to be built in the U.S. and other parts of the world. But opportunity is neither certainty nor necessity. A host of challenging problems and issues must be addressed before a new generation of plants will be built in the United States. The Department of Energy is working cooperatively with the nuclear industry and others to address these concerns.

CURRENT NUCLEAR SITUATION

World electricity generation from commercial nuclear power plants expanded significantly in the 1970s and early 1980s. For most countries of the world, and in the U.S. in particular, nuclear power plant construction increased in response to the oil shortages of the early 1970's. In 1970, the world's nuclear plants provided only 3 percent of the total electricity generated. By 1980, nuclear generated 12 percent of the world total, and in 1985, nuclear's contribution rose to 25 percent. Since then the nuclear share of the total electricity generation has decreased to about 23 percent. At the end of 1994, there were 432 nuclear power plants operating in over 30 countries around the world, with a total electrical capacity of 340.7 net gigawatts electric (GWe). In the U.S. today, nuclear power provides 22 percent of the utility-generated electricity, or the equivalent of about 99 GWe.

Table 1: World's Leading Nuclear Power Producers

Country	Total Capacity (Gwe)	% of World Total Nuclear Production
United States	99.1	29.1%
France	58.5	17.2
Japan	38.9	11.4
Germany	22.7	6.7
Russia	19.8	5.8
Canada	15.8	4.6
Ukraine	12.7	3.7

(Source: U.S. Department of Energy)

Nuclear power programs in many countries around the world slowed during the late 1980s and leveled off in the early 1990s. World nuclear growth is projected to remain fairly level through 2010 at an annual growth rate of 0.2 percent. Use of natural gas is expected to increase while projections show nuclear generation shrinking as the number of reactors retired surpasses the number of new nuclear plants being commissioned. As a result, the nuclear share of power generation worldwide is expected to decline from 17 percent in 1992 to 14 percent by 2010.

The cost overruns, delays and cancellations that characterized many U.S. nuclear power plant programs linger in the minds of utility executives, financiers and the public. While the Watts Bar 1 plant is close to coming on-line, it is a full decade late in doing so, at a cost to TVA of approximately $7 billion. And while costs are now coming down, nuclear plants -- once considered worth their relatively large up-front costs because they were projected to be inexpensive to operate -- became more expensive to operate than coal-fired plants in 1986. Utilities, in a new and uncertain era of competition, will not choose large cost uncertainty and potentially long delays.

This is not to infer that nuclear utilities are not implementing actions to ensure cost competitiveness of their operating nuclear plants. In fact, utilities have taken significant steps to improve plant performance. Most notable is the increase in average plant capacity factor from 58.5 percent in 1980 to 65.1 percent in 1988, and to 75.1 percent in 1994, a rise of 17 percentage points. In addition, the number of nuclear plants with improving capacity factors continues to increase.

Table 2: U.S. Nuclear Plant Capacity Factors

	1988	1991	1994
Average Capacity Factor	65.1%	70.2%	75.1%
No. of Units with Capacity Factors > 70%	45	55	78
No. of Units with Capacity Factors > 80%	15	27	47

(Source: U.S. Department of Energy)

Utilities are also reducing the amount of time nuclear plants are out of service. Outage time for maintenance, repairs and refueling activities has continued to decrease in recent years from an average outage duration of 78 days in 1990 to an average of 55 days in 1994. These statistics illustrate significant strides by the utilities to improve nuclear plant performance and ensure continued cost-competitiveness.

These high costs, linked with an unexpected leveling of electricity demand in the 1970s, set the stage for slow growth in nuclear power. Electricity demand in the U.S. was growing at about 7 percent per year during the 1960s and early 1970s, doubling every 10 to 12 years. Following the oil embargo of the early 1970s, the related sharp increase in interest rates, and the resultant economic recession, U.S. electricity demand decreased to approximately 2.0 percent per year. This dramatic decrease in electricity demand forced utilities to stop new plant orders, cancel existing plant orders and, in some cases, stop construction on existing plants. Thus, the primary reason for the lack of new nuclear plants was a basic business decision by the utilities that there was little need for more baseload generating capacity.

Growing public concerns with nuclear technology following the Three Mile Island and Chornobyl accidents brought programs in the U.S. and Europe to a near halt. Now, almost ten years after Chornobyl, if the general public is less concerned about nuclear accidents, they are still skeptical about the wisdom of building new nuclear plants.

In the United States, the advent of utility deregulation and the potential for generation unbundling will make the task of nuclear plant marketeers even harder in the meantime -- especially when integrated resource planning continues to push utilities to implement energy efficiency programs and build natural gas-fired electric generating capacity. Moreover, while the NRC has made substantial progress toward establishing a new design certification rule, it is clear that much remains to be done to assure potential customers that the nuclear plant licensing process in the U.S. is stable and predictable enough to avoid the costly delays of the past. Finally, there is the problem of spent fuel and low-level radioactive waste management, which will need resolution before nuclear plants will be seen as a realistic option in the future.

ENERGY OUTLOOK

The U.S. Department of Energy (DOE) expects that electricity will remain the fastest growing form of end-use energy worldwide through the year 2010. Electricity is the most efficient and versatile means of applying energy, and world energy growth is expected at an annual rate of 2 percent for the period 1990 to 2010. Although electricity demand growth in OECD countries has slowed to a rate of 1.7 percent, strong growth is expected in non-OECD countries at an annual rate of 2.3 percent, with China, South Korea, Indonesia, Singapore, Thailand, Phillippines and Taiwan experiencing the largest increases. This growth is due to the aggressive industrialization and urbanization in these areas as well as energy independence policies. World-wide electricity consumption is expected to grow from 10.4 billion kilowatt-hours (bkWh) in 1990 to 15.3 bkWh in 2010.

While some portions of the world are experiencing strong growth, the U.S. can be expected to pursue energy efficiency aggressively as a means of mitigating the need for new capacity. Industry is confident that cost-effective efficiency measures -- many of which will be installed by utilities as they become full-service energy companies rather than simply producers and deliverers of energy -- will be an important part of the energy picture for many years to come. These measures will result in moderate growth in electricity consumption. U.S. electricity demand is projected to grow at an annual rate of 1.3 percent per year for the period 1990 to 2010; this translates to growth from 2.7 bkWh in 1990 to 3.5 bkWh in 2010.

The Department's Energy Information Administration believes that approximately 140 gigawatts of new electric generating capacity will be required in the United States by 2010. Most of this growth is expected to be satisfied by new fossil-fuel power plants and implementation of about 16 gigawatts of cogeneration and 20 gigawatts of renewables. Neither the Department nor industry expects new nuclear plants to come on line during this period. In the near to intermediate term, new additions to the U.S. grid are expected to be fueled primarily by fossil fuels, cogeneration and some application of wind, solar and other renewables. The opportunity for new U.S. nuclear plants arises when consideration is given to increased demand in the post-2010 period.

INTERNATIONAL NUCLEAR OUTLOOK

Even if the United States will not build plants for another decade or more, 20 countries have a total of 98 nuclear units under various stages of construction, and many of these countries plan to continue building nuclear plants into the future. It is important to note that much of the interest in nuclear energy in the intermediate term is in countries with high-growth economies that are concerned with the same issues of energy supply and energy independence that led the U.S. to build its current fleet of plants. Further, many countries -- citing the environmental problems of Eastern Europe and the Former Soviet Union -- see nuclear power as a way of maintaining and expanding rapid industrial growth while preserving environmental quality.

Asia

In 1994, the only new nuclear capacity to come on line in the world -- six plants providing over 5000 MWe of electric capacity -- entered service in Asia. Japan, the Korean peninsula, the People's Republic of China, Phillippines, and Taiwan account for 16 percent of the world's total nuclear capacity. Most of these countries (except China) lack indigenous resources and have developed nuclear programs in an effort to become energy independent. In many of these countries, the cost of nuclear energy programs is secondary to the promise of future freedom from supply interruptions. Currently, there are 68 units totaling 54.0 Gwe operating in these countries, with an additional 34.4 GWe of new capacity from 37 plants under construction. By 2010, East Asia is expected to have approximately 100 GWe of nuclear capacity on line.

Japan has a well-established and ambitious nuclear program with 49 units totaling 38.9 GWe at the end of 1994. Japan's nuclear expansion plans will increase capacity to between 50 and 55 GWe by the year 2010. This expansion has 18 units identified, totaling 20.1 GWe, in various stages of construction. Japan's long-term plans include the construction and operation of reprocessing and recycling facilities to handle nuclear waste domestically.

South Korea is the second largest nuclear energy producer in the Pacific Far East, with 10 units operating totaling 8.2 GWe. Eleven nuclear construction projects are under construction and will more than double the current capacity by the year 2015.

China did not have a nuclear power plant until 1991. During 1995, its third unit, Guangdong 2, went into commercial operation. China has initiated an aggressive nuclear expansion and plans on having a nuclear generating capacity of 50 GWe by the year 2020. China's plan will utilize international vendors as well as its own Pressurized Water Reactor design. China has contracted with Framatome for two 985 MWe reactors for units at Daya Bay and with Westinghouse for two steam turbines at the Qinshan units 2&3. China is also planning on nuclear independence by developing spent fuel reprocessing and plutonium recycle.

Europe

Western Europe has 10 countries with a total capacity of 121.9 GWe from 151 nuclear units. This accounts for 43 percent of the total electricity generated in the region and 36 percent of the world's nuclear capacity. Many countries are predicting increased electrical demand in the future but, with the exception of France, nuclear power is not the fuel of choice. A total of 10 units are under construction, 8 of which are in France. As in the U.S., economics, public concerns about safety, and radioactive waste issues have led to an uncertain future in western Europe for nuclear energy.

Eastern Europe, Ukraine and Russia have 65 units in the region generating a total of 43.5 MWe at the end of 1994. Growth in Eastern Europe is aggressive, with an increase of 22 GWe by 30 units currently under construction. The majority of the growth is in Russia and the Ukraine, with 10 and 7 new units, respectively, planned. These countries are in a difficult transition from centralized to market economies and will require significant international financing to carry out their plans.

Other Countries

Other nuclear growth areas in the world include India, which currently has 9 operating nuclear units with a generation capacity of 1.5 GWe. India plans to complete an additional 8 units which are under construction. One plant, Kakrapar 2, is scheduled to come on line sometime this year. Finally, Mexico recently started its Laguna Verde 2 reactor this year, giving that country its second operating reactor and 1.3 GWe of nuclear capacity.

THE FUTURE OF NUCLEAR ENERGY IN THE U.S.

As we mentioned earlier, the opportunity to build new nuclear power plants in the United States will probably not materialize until the time comes to construct plants that would begin operation after 2010. This means that at least another ten years will pass before the next U.S. nuclear plant is likely to be ordered. Such a hiatus reflects the ability of U.S. utilities to turn to cheap, abundant alternatives such as coal and natural gas to fuel electricity generation, and to make that generation more efficient.

The secondary effects of another ten years or more before the next plant is ordered are very serious and of considerable concern to the Department, electric utilities, and the U.S. nuclear industry. These effects are already being felt. The main impact is the slow erosion of technical expertise and industrial capability. As a result, the U.S. is becoming less competitive with each passing year and is losing its long-held world leadership in nuclear technology to countries such as Japan and France that have more aggressive nuclear programs. Major corporations such as Westinghouse, General Electric, ABB-Combustion Engineering, Babcock & Wilcox -- which were once giants of the nuclear industry -- have dramatically scaled back or eliminated their nuclear programs, closed down their manufacturing facilities and laid-off or reassigned key personnel. Other nuclear entities such as architect-engineering and service corporations have turned their attention to other energy sectors.

Perhaps most importantly, this erosion is being felt at U.S. universities and colleges. Many of them can no longer sustain nuclear engineering departments to educate and train the young engineers and scientists needed to build next-generation plants. Some have told us that they would shut their research reactors down immediately if they could afford the costs associated with decommissioning and decontamination. The Federal Government, which provides some funding to these schools, is not doing enough. Congress provided the Department only about half of the $6 million which it requested to assist university nuclear engineering programs and to provide fuel and upgrades to university reactors. The nuclear utilities -- who have the most to lose -- also provide some assistance. Many U.S. utilities provide funds to our universities through a cooperative program with the Department of Energy. But industry isn't doing enough. In some, very disturbing cases, universities receive no funds at all from local nuclear utilities in their states. Therefore, we would encourage all participants in the U.S. nuclear industry to take an active interest in preserving this country's nuclear engineering education base.

Why Nuclear Is Important for the Future

The Department of Energy believes that preserving the nuclear option is important to the energy future of the U.S. It is in our national interest to maintain fuel diversity in electrical generation. Diversity helps ensure continuity of power when unforeseen perturbations occur in one or more fuel sources. Two winters ago in the U.S., with coal barges frozen in the rivers, many utilities found that nuclear power was the only entirely reliable source of electric generation available to them.

The United States has important economic and environmental interests in finding alternatives to the burning of fossil fuels. While energy reserves remain significant -- 43 years worth of oil, 60 years of natural gas, and 235 years of coal are believed to exist in proven fields throughout the world -- the cost of obtaining and delivering these energy sources is likely to increase over time. This is particularly true in the case of natural gas; use of gas for electricity generation is predicted to increase by more than 60 percent by 2010. Further, the greater application of nuclear energy could help mitigate the environmental impacts of burning fossil fuels on an industrial scale.

We noted earlier that Asia is expected to be the largest market for nuclear power plants in the world over the next fifteen years. Nuclear energy is often a desirable option for nations without abundant natural resources or the infrastructure to deliver those resources. The Department will work with many such countries to develop their renewable resources and maximize the use of energy efficiency. However, it is clear that rapid industrialization requires the construction of baseload power plants. For several countries -- most notably China -- this will include nuclear power plants. In this case, the U.S. has a vital interest in helping industrializing countries that choose the nuclear option to build the safest plants possible. The world community has a vital interest in assuring that the nuclear safety problems that arose in the Former Soviet Union are not replicated in China.

THE FUTURE OPPORTUNITY

The future of nuclear energy depends upon the resolution of several issues, including the following:

The Cost-Competitiveness of Nuclear Energy

We expect that nuclear energy can be competitive with other energy options. This revived competitiveness will depend on how well industry learned the lessons of the past and on how effective the new NRC licensing process proves to be. With the cost of natural gas expected to increase and the added cost of scrubbing emissions from coal-fired plants, standardized, certified nuclear power plant designs should be highly competitive in the next century.

Current plants must pave the way by continuing to reduce operating and maintenance costs and further increasing capacity factors. Much progress has been achieved on both counts, and industry is to be congratulated for its efforts. But if a new generation of plants is to be constructed, even more must be done.

Nuclear Waste Management and Disposal

The Department of Energy is responsible for the disposition of U.S. commercial spent fuel. DOE collects a fee (1 mill per kilowatt-hour of nuclear-generated electricity) from the utilities to fund the activities of the Office of Civilian Radioactive Waste Management. Clearly, the Department has not performed up to expectations in the waste management area, but this is largely because the expectations for the program, set down in law in 1987, were never realistic. The Department has now set the waste program on a realistic track and expects to complete a viability assessment of the Yucca Mountain site in 1998.

The Department has no programmatic responsibility for low-level waste disposal. The establishment of regional disposal sites is being addressed by the States through regional compacts. This process has taken longer than we anticipated when the Low-Level Radioactive Waste Amendments Act passed in 1987. Fortunately, volume reduction technology has made it easier for utilities and other generators to store waste materials on-site. Nevertheless, it is important that great emphasis be placed on moving the low-level waste siting process quickly forward.

Increasing Nuclear Trade

The importance to industry and the Government of nuclear exports should be apparent from the issues which we have raised so far. Without strong domestic demand for nuclear power plants before 2010, profits from abroad become the key to retaining our technological leadership and invigorating our technical base. The Department of Energy understands and supports this. Most important in this regard has been our cooperation with industry on the development of the new generation of light water reactors. Equally significant has been the Department's efforts to establish an international nuclear liability regime that would allow U.S. companies to work abroad with the same sort of protection that Price-Anderson provides in the United States.

In the long-term, the Department's most important contribution to expanding nuclear exports may be our engaging in dialogue with those countries with the greatest potential nuclear markets. This can be a frustrating effort that takes a long time to bear fruit, as many of the companies that work closely with us can attest. The Department has opened doors to communication and cooperation that previously remained shut in such markets as China and India, and has set a direction for the Department to continue this strategy. Much remains to be done, but the Department is committed to the proposition that expanded dialogue and collaboration is the most promising path toward our common goals of improving nuclear safety and nonproliferation worldwide while strengthening our industry through increased nuclear exports.

Improving International Nuclear Safety

We must work to ensure the safety of nuclear power plants throughout the world. In the U.S., NRC regulation and Institute of Nuclear Power Operations (INPO) operations and training standards continue to set the standards for the rest of the world. This same level of standards has not been utilized by the rest of the world in the development of their nuclear programs. Of particular concern are the operating Soviet-designed reactors in the Former Soviet Union and Eastern Europe. In addition, the safety of the existing plants in the U.S. needs to be assured as these plants age.

In 1986, the dramatic accident at the Chornobyl nuclear power plant in Ukraine focused international attention on Soviet-designed nuclear power plants, which currently operate in Central and Eastern Europe and the New Independent States. The Department works cooperatively with these countries, other countries with advanced nuclear power programs, and international organizations to improve the safety of operating Soviet-designed nuclear power plants. The United States' program originated with commitments made in 1992 involving the older Soviet-designed power plants. Our efforts, combined with those of others in the international community, have contributed significantly to an improved nuclear safety awareness and culture in many of the countries operating Soviet-designed nuclear power plants.

For example, in Russia and Ukraine, as part of our cooperative activities, advanced nuclear training centers have been established to assist those countries in developing and implementing modern staff training programs. At the Ignalina nuclear power plant in Lithuania, we have contributed to the first-ever probabilistic risk assessment of an RBMK-type reactor. The information gained from this risk assessment is now being used in the development of emergency operating instructions for the Ignalina plant, and will contribute to other safety improvements.

The work that the United States is performing to help establish a nuclear safety culture and infrastructure in the countries that operate Soviet-designed nuclear power plants is vital to our own national interests for several reasons. First, the stability of these emerging democracies would be threatened by the social, economic and environmental impacts of another serious nuclear accident, such as the one that occurred at Chornobyl. Second, a serious accident could have many direct environmental and economic impacts on our European allies. Third, a severe nuclear accident could dramatically affect the viability of nuclear power throughout the world. And finally, a serious nuclear accident could place at risk rapidly expanding U.S. commercial investments in these countries.

Over the next few years, the Department of Energy's program will support these countries in acquiring modern safety technology and methods that will enable them to participate as equal partners in the international nuclear community. Russia, Ukraine, and the Central and Eastern European countries increasingly will have much improved capabilities to make and implement independent, well-founded decisions concerning the continued operation or upgrade of existing reactors based on their own nuclear safety standards.

Completing Development of Advanced Light Water Reactors

In meeting the demands of the utility industry, the NRC, and the American public to ensure that the next generation of nuclear plants in the U.S. is even more safe than the first generation, nuclear planners have worked hard to develop advanced reactor designs which are simpler, more robust, and less subject to damage from any possible upsets in their systems. The Electric Power Research Institute published a Utilities Requirements Document in 1990, reviewed by industry and approved by NRC, which specified the stringent safety requirements for both the large (1200-1350 MWe) evolutionary designs and the smaller (600 MWe) advanced passive designs.

As a result, the evolutionary designs (Advanced Boiling Water Reactor [ABWR] and System 80+) feature simplified designs, increased design margins, increased redundancy in emergency core cooling systems, incorporation of human factors in design, and improved constructability and maintainability. These improvements have reduced core damage frequency probabilities and lowered expected worker radiation exposures, contributing to increased safety margins. The mid-size advanced passive plant designs (Simplified Boiling Water Reactor [SBWR] and AP600) have greatly increased safety margins due to the reduction in number of components in the plants, and the use of passive safety features, such as natural circulation and gravity flow emergency cooling systems, to prevent and mitigate potential accidents. The increased safety margins result in a reduced dependence on immediate operator actions to protect the reactor in the event of an accident.

The Department of Energy's Advanced Light Water Reactor (ALWR) Program is a cost-shared program with the utility and nuclear industries. Industry is actively contributing greater than 50 percent of the costs associated with the ALWR programs. The ALWR program has produced two evolutionary plant designs which have received final design approval by the NRC and are in the final stages of certification in the U.S. The ABWR is already being built in Japan. The advanced passive plant designs are also moving ahead towards certification. Final design approval of the AP-600 nuclear power plant is expected in 1996.

The ALWR standardized designs are being certified under the 10 CFR part 52 Combined Construction/Operating License regulation in order to prove the process. The NRC is currently in the rulemaking process for the ABWR and System 80+ designs. This is a public process, with the nuclear industry, DOE, and the public all providing comments. The NRC is currently evaluating input from DOE and industry that could further simplify the rulemaking process and help avoid lengthy hearings for some future plant design changes. It is anticipated that the rulemaking process will be completed in early 1996, resulting in final design certification for these two plants.

The ALWR program participants are also sharing costs with industry and DOE in developing two of the ALWR designs further. The ABWR and AP-600 First-of-a-Kind Engineering (FOAKE) programs are intended to complete the standardized designs to the point where construction costs and schedules can be assured prior to utility commitment, as well as develop improved construction techniques.

Assuring the Safe, Continued Operation of Current Plants

We must make sure that our current generation of operating nuclear power plants continue to operate at maximum efficiency and safety for their useful lifetime. The operating nuclear power plants play a critical role in ensuring fuel diversity. However, by the year 2015 U.S. nuclear plants accounting for approximately half of the installed nuclear power base will have had their current operating licenses expire; by 2033, all existing nuclear plant licenses will have expired. The Department is addressing, with industry participation, significant generic technical issues facing these nuclear plants that affect their useful life, performance and license extension.

The major DOE programmatic activities in this area include: support for industry efforts to resolve major technical issues and demonstrate the NRC's license renewal rule; a reactor pressure vessel annealing project to demonstrate the viability and effectiveness of the annealing process; a project to develop a process to mitigate degradation of reactor vessel internal components; and research into prevention of cable degradation.

CONCLUSION

The Department of Energy is working to assure that nuclear power has a safe future in the United States and elsewhere in the world. The opportunities for nuclear are close at hand and it is very possible that the next U.S. plant could be ordered within the next decade. Building the next generation of plants will require leadership from all the involved parties. Electric utilities in particular, if they want to maintain the nuclear option, must be vocal and visible in their support for nuclear technology and education. They must be willing to commit their time, funds and people to work with government, the public and other industries to see new plants built.

But before any new plants are built, government and industry must work together to resolve important issues impacting the viability of nuclear energy. It will be vitally important to take all the concerns of the public seriously and be aggressive in addressing them. Building the next generation of plants will require keeping the current generation safe and cost-effective. Clearly, a major nuclear accident anywhere will claim as one of its victims the prospects for a revival in the U.S. nuclear industry.

And thus, if nuclear power is to play a part in the energy picture of the twenty-first century, we must first be certain that we have learned to correct the mistakes of the past.

A VISION OF THE SECOND FIFTY YEARS OF NUCLEAR TECHNOLOGY

Don W. Miller

Professor and Chair
Nuclear Engineering
The Ohio State University

Vice President/President-Elect
The American Nuclear Society

INTRODUCTION

In my remarks today, I will consider the following topics:

- International Nuclear Societies Council Vision for the Second Fifty Years of Nuclear Energy [1,2]
- Health Effects of Low Level Radiation[3]
- Responsibility of Nuclear Technologists

INTERNATIONAL NUCLEAR SOCIETIES COUNCIL VISION FOR THE SECOND FIFTY YEARS OF NUCLEAR ENERGY[1,2]

The most compelling issue the world will face in the next 50 years is the quality of life in the poorer regions of the world. Quality of life in the face of population explosion. Quality of life that is dependent on the three essential and intertwined elements of adequate energy, food supply and medical care.

Our vision of the future must be set in that context.

Although nuclear technology can and will have a major role in all these elements, I will direct my initial comments to the most complex element, energy, and therefore, nuclear power.

Global energy demand is being driven upward by a number of factors:

1. World population is increasing by nearly 100 million each year. mainly in the developing countries. Experts tell us that the world's population will double by the middle of the twenty-first century, to over 10 billion people, and nothing will stop that doubling, short of a disastrous global epidemic affecting hundreds of millions of people.

2. Television and the rapidly evolving internet already bring images and information to and from all parts of the world. These revolutions in communication are and will generate global expectations for a better or a different quality of life.

3. Quality of life depends strongly on societal wealth and wealth is created by efficient use of resources and energy. The wealthy countries use large amounts of energy and their citizens today are healthier, better educated and live longer than any other people at any other time in the history of civilization.

4. Although the environment currently may not be a high priority in most of the world today, its fundamental importance will eventually dictate the way of life world wide.

Energy is the key to overcoming the scourges of hunger, disease and poverty - it's essential to generating wealth and it's essential to protecting the environment in a heavily populated world.

We can look at energy demand in the next 50 years from two points of view.

The first is based on improving the quality of life in the world at large. On a per capita basis, people in the developing countries use approximately one tenth of the energy used by people in the rest of the world. There are a number of indicators suggesting that their quality of life would be substantially improved if their energy consumption were to increase by five to 100 Giga Joules (GJ) per capita. That would bring them to about half of the current use in the United States and Canada. If the future world consumed 100 GJ per capita then the global energy demand would approach 1000 EJ (E+21 Joules).

A second point of view is historical where energy demand has increased at 2.3% per year for the past 130 years. If that growth were to be continued over the next half century, the global energy demand would be projected to exceed 1000 EJ.

Thus both these approaches suggest a global annual energy demand of about 1000 EJ by 2050 - a projection close to the World Energy Council projection[4], which is three times today's global use rate.

We do not foresee any fundamental limitation that would prevent such energy use. The planet can easily support 10 billion people if there is a will at the political level, and if there is adequate energy and food supply.

On the issue of food supply, a key requirement will be distribution, since the issue of adequate availability is not, nor will it in the future be a limiting factor.

On the issue of adequate energy, this requirement can be met if the current high energy consumers share their technology, do not limit increase in energy use in the developing countries, and curtail their own rate of increase in energy use.

Marchetti[5] has shown us that the normal pattern is for energy sources to grow in use until new sources displace them. The market share held by coal and oil is already declining, while natural gas and nuclear are rising on a world-wide basis.

Marchetti's model suggests that gas and nuclear will be the dominant energy sources in the next 50 years, with gas accounting for about 60 %, coal and oil together 6% and nuclear over 30 %. That estimate may be extreme since it would require that gas reserves increase by an order of magnitude.

Natural gas reserves could be extended by limiting their use for base load generation of electricity, and by concentrating on more thermally efficient "end uses" such as heating, air conditioning and industrial process heat.

If today's mix of energy uses were maintained, the future demand would be a three fold increase use of nuclear over the next 50 years.

If Marchetti's model proves correct then the demand for nuclear energy will be increased by an order of magnitude.

Finally, if the Rio Accord on limiting carbon dioxide emissions becomes accepted globally, then the demand for nuclear can only be met by a 20 fold increase in nuclear electric generation!

We know of no fundamental limitation in nuclear fuel resources. Uranium and Thorium are both abundant in the earth's crust and oceans, and with current technology, fission energy can conceptually supply the future demand for thousands of years.

So, how large a nuclear power supply is practicable? Take, for example, nuclear energy supplying one-third of the energy by the year 2050. This would require a nuclear capacity of about 5000 GWe and would require construction of 100 GWe of new capacity per year over the next 50 years. This 100 plant/year rate is substantially more than the maximum rate of construction achieved to date, but it is achievable from an industrial point of view. The availability of capital and political willingness will likely be more important factors rather than technology or industrial capability.

The key factor in an expanding global economy will be economics. Wherever natural gas is available in the pipeline, that will be the economic competition over the next one or two decades. So how will current nuclear designs compete with low cost natural gas? Because of high capital costs, current designs will not compete with low cost pipeline natural gas in the next decade or two.

Although we do not believe it is good policy to use natural gas for base load electric generation, especially in countries where other resources and the technology to use them effectively are available, we realize nuclear will have to be competitive with natural gas. Before we consider changes required for nuclear energy to become competitive, we need to consider the current status.

Many of the current operating plants in the US were completed at substantially higher capital cost than anticipated as a consequence of extensive delays and changing regulatory requirements. Added regulatory requirements plus changes in designs also lead to rapidly increased operating costs.

In the past few years operating costs have begun to stabilize and plant capacity factors have increased by more than 25%. In the regulatory arena we have seen changes in licensing requirements and the introduction of risk based regulation. The revised licensing requirements are expected to result in reduced plant construction times and risk based regulation will lead to more efficient use of resources.

Four standard plant designs, two evolutionary and two passive, have resulted from the US Advanced Light Water Reactor Program. All these reactors are expected to have lower capital and operating costs. The evolutionary designs have been certified by the NRC, and one of the passive designs should receive certification in the next year or two. Recently, construction of one of the certified designs was completed in Japan in less than five years.

Although there have been definite changes both in plant designs and in the regulatory process over the past several years, there is more to be done to further reduce both capital and operating costs. The key changes will be modular designs and production line manufacturing. These changes will be required for nuclear energy to be competitive with low cost pipeline gas and to assure that nuclear energy meets its expected and needed contribution to the world wide energy mix.

There are other prerequisites for an expansion of nuclear power.

First, although it is unlikely that there will be another major accident on the scale of Chernobyl, another accident of this scale would re-kindle public fear and would seriously decrease prospects for nuclear technology to fulfill its promise of abundant energy for the world. Severe accidents must be prevented.

Second, the public needs assurance that everything is in place to handle the radioactive waste from uranium mining, from operation and decommissioning of current reactors and fuel cycle facilities, and from all other uses of radioactive substances. The topic of radioactive waste is an emotional issue that is central to the future of nuclear energy as a sustainable technology. Fortunately, in recent decades there has been much progress in many countries towards practical, economic and safe disposal of radioactive waste. The issue will be resolved through the cooperative and determined efforts of the industry, regulators and the political sector.

The use of nuclear technology for energy production and medical therapy and diagnostics has a history exceeding 30 year and a proven record of success. In the next 50 years one of the most important uses of nuclear radiation will be in the preservation of food.

Sterilization of food by irradiation replaces the traditional use of temperature extremes-- cold and hot-- and extreme drying techniques. The capability of providing for long-term food storage without decay has the potential to substantially increase the ability to transport food long distances, and thus increase food availability worldwide.

Malnutrition is rampant in many parts of the world today and stems not only from marginal food production, but also from the fact that a large fraction of food becomes infested and lost to bacteria and insects. Eliminating this wastage by using radiation could effectively double the availability of food from current production. The technology is available, but the process is not yet used widely, even in the industrialized countries because of public concerns. As its benefits and safety become more widely accepted we project that food preservation by irradiation will be gain general use worldwide over the next fifty years.

HEALTH EFFECTS OF LOW LEVEL RADIATION[3]

I will now shift my remarks from specific applications of nuclear technology to an issue important to all applications of nuclear technology, the health effects of low-level radiation.

In our country, we are subject to a paranoia about radiation that is fueled by self-interest groups whose real agenda often has more to do with social change than public protection. So what are the effects and even benefits of low-level radiation

There are numerous relevant dose-response data, organized into significantly exposed populations, including:

- •Japanese survivors of the atomic bombings of Hiroshima and Nagasaki.
- •Occupationally exposed populations (including radiologists and other medical practitioners).
- •Medically exposed patient populations.
- •Radium body-burden populations.
- •Weapons and facilities release populations (with military weapons tests observers).
- •High natural background radiation-exposed populations.

The data on health effects from low-to-moderate radiation doses in these populations show no adverse health effects at doses below about 20 cGy (1 cGy equals 1 rad). (The exception is for moderate doses, above about 5 cGy, of high dose rate X rays to the fetus at cell differentiation during the second trimester of development.) In addition, data show that adverse health effects are small for doses in the range of 200-400 cGy at low dose rates and for fractional high dose rate exposures.

Statistically significant data actually show below-normal adverse health effects (i.e., health benefits) at low-to-moderate doses compared with unexposed populations.

In conclusion there is no statistical evidence of adverse health effects from radiation doses of less than 20 rads or radiation dose levels substantially above current regulatory limits.

Discussion of the Data

From the time of Ebon Byers' death in 1932, radiation protection policies and research have increasingly fostered public fears and supported costly government regulations and programs. These costs have been largely borne with minimal questioning, directly by the general public, due to the ability of affected governments, utilities, and private corporations to readily pass these costs through increased medical and utility bills with small adverse competitive costs. These large costs might have been wasteful, but were moderate compared with the primary costs of nuclear technology operations and the cost of providing necessary protection for workers and the public from hazardous radiation exposures.

Recently, however, these radiation protection excesses have resulted in large incremental public costs, with even more proposed, with no accompanying public health benefit. Currently, these policies especially affect radioactive waste management and site decommissioning costs. The excessive costs are reducing the viability and public benefits of many nuclear technology applications, and humanity is losing major advances and contributions to human health and well-being, without benefit to public health.

Current work in Japan, and similar work in China, with the beginnings of small efforts in Europe and elsewhere, along with some US. work in biology performed in cancer and genetics research, may be the hope of the future. Meanwhile US. policy and public progress in the use and reasonable control of radiation and nuclear technology applications stagnate in favor of fostering public fear and unjustified public costs.

Current Federal agency rule makings propose to still further reduce public radiation dose limits and increase public costs, with no public health benefits. This is true even when arbitrarily applying the linear model dose-response, since the proposed regulations would reduce exposures that are "less than 1 percent" of public radiation doses to limits that are "much less than 1 percent" of public doses; and to "protect" individuals from doses that are small compared with natural variations in daily living.

We as knowledgeable and credible nuclear technologists must work to change these policies. There is a need to require the federal agencies to address the factual evidence on radiation dose-response health effects that are necessary to establish valid regulatory standards to assure public health and safety; and to constrain ever-increasing regulatory and program costs that provide no public health and safety benefit.

The linear theory of radiation damage, which assumes that any radiation is harmful, however small, and even below backgrounds that we live in on a daily basis, is a very costly theory, a theory supported neither by scientific evidence nor by plain old fashioned common sense.

CONCLUDING COMMENTS

Consider that less than 100 years ago the fundamental science on which nuclear technology is based was essentially unknown. Bohr was formulating the fundamental understanding of the nucleus, Einstein had not published his well known mass energy equivalence and it would be more than 30 years before the fundamentals of fission and fusion were known. It was only 53 years ago when Fermi demonstrated sustained fission.

Today, nuclear technology contributes to the treatment of more than 25 percent all who are admitted to hospital.

Nuclear power now supplies more than 20 per cent of U.S. electric energy and 7% of the world's energy.

Looking to the needs of the world over the next 50 years, we see that an even more important contribution will be demanded of nuclear technology, not only in the production of energy, but in many other uses that will benefit humankind.

Nuclear technology has more potential for solving world hunger and energy problems than any other single technology known today.

No other technology has so much potential to reduce the real risks of so many people and no other technology is as demanding of perfection in its use as is nuclear technology. We have demonstrated by operation of the more than 100 power reactors in this country over the past ten year that we can safely meet this demand.

RESPONSIBILITY OF NUCLEAR TECHNOLOGISTS

What is the responsibility of the nuclear technologist?

We as professional nuclear technologists, who have the knowledge and expertise in the application of nuclear science and technology, have a morale obligation --- an ethical mandate -- to assure that nuclear technology is available for use by the underprivileged majority as a means to substantially reduce their real risks--- risks arising from lack of food and lack of energy.

We must provide the leadership[p to change the regulation of nuclear technology such that it properly reflects current knowledge and understanding of the health effects of low-level radiation.

We must design standardized and easily manufactured nuclear power plants. Plants that can be constructed at a competitive capital cost, and will operate with safety and reliability performance characteristics equal or superior to current plants, but at substantially reduced costs.

We must continue our efforts to educate the public on the benefits of preservation of food by radiation and the generation of energy by nuclear fission.

We as knowledgeable professionals can not stand by and permit a minority -- a minority subject to minimal risk in their own daily lives whether their reasons be lack of knowledge, fear of real or perceived risk, financial or self serving, or simply the desire to maintain the status quo at the expense of others -- to inhibit the use by the underprivileged majority of a technology that will substantially reduce their real risks and improve their quality of life.

Our vision of nuclear technology is a vision of hope for the world.

REFERENCES

1. S.R. Hatcher, "International Nuclear Societies Council Vision for the Next Fifty Years of Nuclear Energy" Presentation to the ANS President's Session ANS Winter Meeting, San Francisco, October 31, 1995.

2. D.W. Miller, Additional Comments (Opinion) and Information on use of natural gas for generation of electricity and current status of the USLWR and ALWR Programs, November 10, 1995.

3. J. Muckerheide, *The Health Effects of Low-Level Radiation: Science, Data, and Corrective Action, Nuclear News*, pp 26-34, September 1995

4. World Energy Council, *Energy for Tomorrow's World*, London, ISBN 0 7494 1117 1, (1993), update in 1995

5. C. Marchetti, *Nuclear Energy and its Future*, Perspectives in Energy, USA 7, pp 19-34, 1992

CHAPTER IV

NUCLEAR POWER GROWTH, NON-PROLIFERATION, AND POLITICS OF NUCLEAR ENERGY

ADVANCED BOILING WATER REACTOR (ABWR) FIRST-OF-A-KIND ENGINEERING (FOAKE)

Roger J. McCandless

Mission and Project Manager, FOAKE
Nuclear Plant Projects
GE Nuclear Energy
San Jose, California

INTRODUCTION

Two Advanced Boiling Water Reactors (ABWRs) being constructed at the Kashiwazaki-Kariwa Nuclear Power Station in Japan represent the latest stages of design evolution in boiling water reactor power plant technology. These units (K-6 and K-7) are rated at 1356 MWe, making them among the largest in the world. These units are being built by Tokyo Electric Power Company (TEPCO), the owner, and by General Electric International, Inc. (GEII, formerly known as GETSCO), Hitachi Ltd. and Toshiba Corporation under a joint venture agreement. The addition of these two newest units will increase the total number of BWRs in TEPCO's nuclear fleet to 17 units with a total electrical capacity of more than 17,000 MWe. K-6 and K-7 are expected to enter commercial operation in 1996 and 1997, respectively. At this writing (January 1996), fuel loading and first synchronization to the grid have been achieved at the K-6 unit.

ABWR First-of-A-Kind-Engineering (FOAKE) is developing the ABWR design for application in the United States. Funding for the ABWR FOAKE Project is provided by GE and its FOAKE associates, including members of the ABWR FOAKE Design Team, the Advanced Reactor Corporation (ARC), representing utility sponsors of the ABWR FOAKE Project, and the United States Department of Energy (DOE). The ABWR FOAKE Project began in June 1993 and is on schedule for completion in September 1996.

The basis for the U.S. ABWR on the FOAKE effort is different than the basis for the ABWRs in Japan. The basis of the ABWR FOAKE design is the ABWR Standard Safety Analysis Report (SSAR) and the Utility Requirements Document (URD). The ABWR SSAR was developed under the ABWR Certification Program in cooperation with the U.S. Department of Energy. The URD was developed under the Electric Power Research Institute's (EPRI) Advanced Light Water Reactor (ALWR) Program. The ABWR SSAR received a Final Design Approval (FDA) from the U.S. Nuclear Regulatory Commission (NRC) in 1994 with NRC Certification expected in 1996.

This paper will describe the process and infrastructure initiatives undertaken on the ABWR FOAKE Project. While advanced design features and advanced technology applied in the ABWR are vital to producing a commercially competitive design with a highly reliable construction schedule and cost estimate, it is equally vital to establish an engineering design, procurement and construction process and infrastructure that will effectively support the ABWR in the marketplace throughout the lifecycle from procurement and construction through startup and 60 years of commercial operation.

MISSION

The ABWR FOAKE Project was undertaken by GE Nuclear Energy to support a key business mission to participate in future nuclear power plant projects worldwide.

The 1995 GE Power Generation Forecast shows substantial electrical power generation capacity additions for the period 1995 through 2004. Worldwide, the the total capacity for this period is projected at 1025 GWe. Asia has nearly half (505GWe) of this projected total. The projections for Europe and the Americas are 321 GWe and 199 GWe, respectively.

Even though the ABWR FOAKE Project is producing a detailed design licensed for U.S. application, it is expected that the ABWR FOAKE design will be suitable for application, with little or no adaptation, in many countries throughout the world. In the near term, countries in Asia are expected to produce the largest marketplace opportunities. In the longer term, substantial opportunities in Europe and the Americas are anticipated.

FOAKE OBJECTIVE

Primary ABWR FOAKE Project objectives are to:

(1) Produce a U.S. Standardized ABWR detailed design.
(2) Produce a construction cost estimate with a high degree of confidence.
(3) Produce a construction schedule with a high degree of confidence.
(4) Perform all design work consistent with the ABWR SSAR and the URD.

In meeting these objectives, the clear aim is to provide the basis for highly competitive ABWR bids worldwide against all alternatives for electrical power generation. Therefore, the design activity focuses not only on achieving low construction costs but on achieving low total energy generation costs, including operation and maintenance costs and fuel cycle costs.

The ABWR FOAKE target construction schedule, from first concrete to commercial operation, is under 48 months. This target will be achieved based on detailed ABWR FOAKE construction schedule work to date. It also appears to be quite realistic based on the construction schedule experience from the two ABWRs nearing completion in Japan. ABWR FOAKE is making use of extensive equipment and civil structural modularization to achieve short construction schedules as has been done in Japan.

Reliable construction cost estimates and schedules depend on both the depth and breadth of the ABWR FOAKE design effort. The ABWR FOAKE design will be about 65% complete at the end of the project, leaving only some of the detailed commodity engineering and site-specific engineering to be completed. Commodity engineering that will remain to be completed post FOAKE includes issuing purchase specifications for bid, equipment vendor selection, modifying all design drawings and documents to reflect vendor-specific equipment details and the the issuing of construction drawings. The scope of the ABWR FOAKE design effort includes the Site Plan, Reactor Building, Control Building, Turbine Island, Radwaste Facility and Service Building. For this scope, the ABWR FOAKE effort is producing detailed P&IDs, logic diagrams, electrical schematics, purchase specifications and equipment data sheets, civil design, including rebar and structural steel drawings, 3D model, including structural and equipment layouts, and the routing of pipes, trays, ducts, cable and wiring. In all, the ABWR FOAKE effort will provide more than 24,000 detailed engineering drawings. This level of effort provides sufficient depth and breadth to support the Project's objectives.

FOAKE PARTNERS

The ABWR FOAKE Project is a $100 Million effort funded 50% by ARC and 50% by GE and the ABWR FOAKE Design Team. ARC's share of the funds is derived from DOE and participating utility contributions.

The ABWR FOAKE participating utilities include Commonwealth Edison Company, Carolina Power and Light Company, General Public Utilities (GPU), Tennessee Valley Authority (TVA), TU Electric Company, Florida Power and Light Company, Duke Power Company, Public Service Electric and Gas (PSEG), Electricite de France (EdF), and Agrupacion Electrica para el Desarrollo Tecnologico Nuclear - A.I.E. (DTN - representing a group of Spanish utilities). Personnel from the Institute of Nuclear Power Operations (INPO) and EPRI joined the ARC and utility personnel in providing lessons learned input from construction, startup and operations to the project through the review of deliverables, design reviews, and topical technical meetings.

Early in the ABWR FOAKE effort a decision was made to carry out the work with a group of companies (the Design Team) who could not only do the ABWR FOAKE design work but who could potentially work together, or with others, in future ventures to deliver ABWR construction projects worldwide.

Several key factors were considered in assembling the ABWR FOAKE Design Team. First, was to involve companies that had substantial and widely acknowledged world class technical experience and resources to support the design, procurement, and construction of nuclear power plants in their areas of expertise and scope. Second, was to select companies with the financial resources to potentially participate in an actual construction project in the area of their anticipated scope. Third, was to select companies that had a long-term interest in delivery of ABWRs worldwide. Fourth, was to select companies that were willing to make financial contributions to the ABWR FOAKE Project. Fifth, was to select companies that were culturally compatible -- i.e., they could work productively and harmoniously together not only on FOAKE but potentially on future ABWR construction projects.

The ABWR FOAKE design team consists of GE, Bechtel Power Corporation (Bechtel), Black and Veatch, Adtechs Corporation - a subsidiary of Japan Gas Company (JGC), Chicago Bridge and Iron Technical Services Company (CB&I), Shimizu Corporation, Hitachi Ltd., Toshiba Corporation, Badan Tenaga Atom Nasional (BATAN- the national atomic eneragy agency of Indonesia), Comision Federal de Electricidad (CFE - the national electric power utility of Mexico), Simulation Systems and Services Technologies Company (S3 Technologies), Equipos Nucleares, S.A. (ENSA), Technatom, S.A., and UTE-INITEC/Empresarios Agrupados (UTE).

GE has overall responsibility for the project and is responsible for the detailed design of the Reactor and Control Buildings, including all of the systems and equipment in those buildings. GE is also responsible for the cost estimate and construction schedule with support from all other participants in their assigned areas of responsibility. GE Power Generation is responsible for the turbine and generator design.

Bechtel provides support to GE in the areas of mechanical and contol and electrical systems design and civil design in the Reactor and Control Buildings. Bechtel is also responsible for the Service Building design.

Black and Veatch is responsible for the Turbine Island design. Black and Veatch also provides the POWRTRAK information management system used on the Project. POWRTRAK (a trademark of Black and Veatch) will be discussed in more detail later.

Adtechs, with support from JGC, is responsible for the design of the Radwaste Facility. Adtechs designed, built and operates the Surrey Radwaste Facility.

Chicago Bridge and Iron is responsible for containment structural steel design.

Shimizu is responsible for the majority of the Reactor Building civil design. Shimizu has major civil construction responsibility on the K-7 Project.

Hitachi and Toshiba will provide consulting and engineering support to GE on the Reactor Building mechanical, control and electrical, and civil design.

BATAN provides support to GE in the areas of mechanical and control and electrical systems design.

CFE provides support to GE and Bechtel in the areas of mechanical and control and electrical systems design and piping stress analysis. CFE also provides support to Adtechs in the area of civil design.

S3 Technologies and Technatom have provided support to GE in the areas of control room and man-machine interface (MMI) design.

ENSA provides support to GE in the area of reactor pressure vessel design, construction schedule, and cost estimate.

UTE provides support to GE and Black and Veatch in the areas of mechancial and control and electrical systems design.

FOAKE INITIATIVES

Standardization

Standardization is a key initiative under the ABWR FOAKE Project to assure competitive costs. A Standardization Plan was developed at the beginning of the project to guide the ABWR FOAKE effort to provide standardization throughout the ABWR plant lifecycle through design, procurement, construction, startup, and commercial operation.

At the top level, standardization is achieved by basing the design on the URD and the SSAR. From the URD, standardization is achieved by designing to a common set of requirements developed under the EPRI ALWR Program. From the SSAR, standardization is achieved by designing to the basis approved by the U.S. NRC under the Final Design Approval (FDA) for the ABWR. This aspect of standardization is reenforced by the fact that the URD has been reviewed by the U.S. NRC and approved through the issuing of a Final Safety Evaluation Report (FSER) for the URD.

The ABWR FOAKE Project engineering is done in accordance with the ABWR FOAKE Project Management, Design and Procurement Manuals which prescribe standardized engineering procedures, codes and standards, methods of calculation, equipment selection, report and drawing contents and formats, as well as a standardized approach to the electronic storage of engineering documents and data. Under the ABWR FOAKE Project, standardization is applied to the engineering of all equipment, buildings and facilities on the site. The clear aim is to provide an extraordinary degree of standardization site-wide for all equipment such as pumps, pipes, valves, heat exchangers, motors, ducts, trays, switch gear, and other equipment, as well as for civil commodities such as rebar, structural steel, equipment supports and pipe hangers.

Procurement

The procurement initiative includes the key elements of standardization: simplification of procurement specifications, establishment of a global equipment supply team, and the use of large procurement packages that cover the procurement of equipment site-wide.

Procurement specifications and equipment data sheets all have standardized contents and formats. The procurement specifications are iterated with equipment suppliers to assure that each specification contains only that information which is required to procure the equipment. The terms and conditions for the procurement packages are also standardized. Iteration with the equipment suppliers helps assure that the technical specifications are compatible with providing off-the-shelf equipment wherever possible and that the terms and conditions are acceptable to the suppliers and are compatible with providing the lowest price quotes. Simplified specifications which contain the minimum information necessary to purchase equipment also contribute to achieving low price quotations because ambiguities and uncertainties are reduced from the supplier's perspective, thus reducing suppliers risk and price.

The use of site-wide procurement packages, wherever possible, also contributes to achieving the lowest possible life cycle costs for the plant through standardization and volume price discounts. For example, certain types of valves are used in the Reactor and Contol Buildings, Turbine Island and Radwaste Facility. A single procurement package for this valve or group of valves will help reduce life cycle costs in several ways: (1) there will be a volume discount from the supplier; (2) the learning curves will be strong for installation and maintenance, thereby reducing costs; and (3) spare parts inventories can be minimized.

Under the standardization program, standardized procurement packages were established early. Equipment suppliers were contacted, agreements were established and two or more equipment suppliers were identified for each procurement package. Most procurement packages have three or four suppliers that are qualified to make quotes. Multiple bids for each procurement package obviously help assure that the lowest costs for equipment are achieved.

Over 80 equipment suppliers worldwide are participating under agreement on the ABWR FOAKE Project. These suppliers are providing quotations for each procurement package that will be used in the final ABWR FOAKE cost estimate.

Global Delivery Team

The members of the ABWR FOAKE Design Team described earlier comprise an ABWR global delivery team capable of supplying ABWRs worldwide. The FOAKE Project provides an environment for the team members to work together to demonstrate their compatibility and to become intimately familiar with the design, procurement, construction, and installation requirements for ABWR delivery. Perhaps most important is that this delivery team is completing the detailed design of the FOAKE ABWR to a high degree of design completion, thus substantially reducing costs related to design uncertainties.

The FOAKE ABWR global delivery team has achieved all of its objectives in producing a detailed and highly standardized design. In future ABWR projects, it is expected that the delivery team may be comprised of some members from the FOAKE delivery team as well as others. The formation of a delivery team for future projects will be done on a case-by-case basis. In any event, the ABWR FOAKE delivery team has set the stage for the delivery of ABWR projects worldwide on the most competitive possible basis.

Information Management System (IMS)

The FOAKE Project set out to perform its work in a highly automated and electronic fashion through the application of an electronic IMS.

At the heart of the ABWR FOAKE IMS is POWRTRAK. Black and Veatch developed POWRTRAK to manage the engineering, procurement and construction processes for electrical power generating stations. POWRTRAK has been widely used with a high degree of technical and commercial success on Black and Veatch's power generation projects for many years. POWRTRAK is maintained and improved regularly to widen and enhance its capability and to keep it current with state-of -the-art software and hardware. GE has worked with Black and Veatch for several years to develop and maintain POWRTRAK for nuclear power plant applications.

POWRTRAK is an integrated engineering database. Each piece of engineering data is stored in one and only one place in the POWRTRAK database. By using POWRTRAK in the engineering process, each engineer has instant access to every piece of engineering data, thereby allowing the engineeering process to proceed with extraordinary speed and efficiency. With POWRTRAK, there is essentially no rework of engineering due to engineers having incorrect or out of date information.

Black and Veatch manages the POWRTRAK database for the ABWR FOAKE Project. A global electronic network has been established allowing every member of the ABWR FOAKE Design Team to benefit from the use of POWRTRAK. Nearly all engineering work done on the ABWR FOAKE Project is done with POWRTRAK, including the development of P&IDs, logic diagrams, electrical schematics, 3D models, routing of pipes, ducts, trays, cables and wiring, etc. The schedule and cost modules of POWRTRAK are used in the management of the engineering schedule, the development of the construction schedule and the preparation of the construction cost estimate.

For electronic document management, the ABWR FOAKE IMS utilizes the Odesta Data Management System (ODMS) to complement POWRTRAK. ODMS allows for the rapid creation, storage and transmittal of text-type documents, as well as drawings. ODMS is used globally by the Design Team to speed document creation, effect reviews, process comments, and assist in the issuing and transmittal process. ODMS is the project central file that is open to all members of the Design Team to allow them to access documents seconds after the documents have been entered into ODMS.

The ABWR IMS is intended to be an integral part of the ABWR product for use through the plant lifecycle through design, procurement, construction, startup, and 60 years of commercial operation. The benefits of an integrated IMS system are well established for engineering, procurement and construction. The engineering benefits have been confirmed on the ABWR FOAKE Project. In the longer term, substantial cost benefits derived from the IMS are confidently projected for ABWR construction projects and operating ABWR plants based on the ABWR FOAKE IMS experience.

Lessons Learned

The ABWR FOAKE Project has had the benefit of lessons learned from the current generation of nuclear power operating plants. Basic sources of lessons learned information for the FOAKE Project include the URD, ARC utility inputs through the ABWR FOAKE document review and design review processes, numerous industry reports (e.g. from EPRI,

INPO, NRC etc.), and databases maintained by the individual Design Team members. Each engineer on the FOAKE Project is responsible for applying lessons learned in the development of the design.

Of course, the genesis of the FOAKE ABWR design is the design of the K6 and K7 units in Japan. The development of those designs incorporated lessons learned in a disciplined manner. In addition, there are lessons learned in the detailed design and construction of K6 and K7, which have benefited the FOAKE design effort.

The lessons learned that have been factored into the ABWR FOAKE design should contribute significantly to increasing performance while reducing costs through the plant life cycle relative to the first generation of nuclear power plants.

FOAKE STATUS

The ABWR FOAKE Project is on schedule for completion in September 1996. As of January 1996, work was about 80% complete. The major work remaining is to complete the issuing of procurement specifications and to finalize the construction schedules and cost estimates.

At this point, our experience indicates that success will have been achieved with the process and infrastructure initiatives undertaken on the ABWR FOAKE Project, including the initiatives on standardization, procurement, creation of a global delivery team, lessons learned and implementation of POWRTRAK and the FOAKE IMS.

SUMMARY AND CONCLUSIONS

1996 is expected to be a very significant year for the ABWR with K6 commercial operation anticipated along with the completion of the U.S. ABWR Certification by the U.S. NRC and the completion of the detailed ABWR FOAKE design. The ABWR team is ready to proceed with further commercialization of the ABWR worldwide.

THE CIVILIAN RECYCLING INDUSTRY: A KEY CONTRIBUTION TO DISARMAMENT

J.A. de Montalembert[1] and M. McMurphy[2]

[1]COGEMA - France
[2]COGEMA Inc. - USA

INTRODUCTION

In a LWR reactor loaded with the usual enriched uranium fuel (the UO_2 fuel), as soon as the nuclear reactions start and during all the core irradiation period (3 or 4 years), some plutonium is produced in situ by U238 transmutation, and part of this plutonium is immediately engaged in fission and thus energy production. At the end, depending upon the discharge burn-up of such fuel, 30 to 40 % of the nuclear power actually comes from plutonium.

This "auto-recycling" was the strong basis for confidence in the introduction of plutonium in fresh fuel (the MOX fuel) where fissile isotopes of plutonium replace the U235 of the usual enriched uranium fuel.

So after reprocessing has separated the components of the spent fuel and properly conditionned the waste, plutonium is recycled as MOX fuel in LWRs (and can also be used in fast reactors).

While in UO_2 spent fuel at current burn-ups the content of plutonium is around 1 %, with 70 % fissile isotopes, a fresh MOX fuel is fabricated with some 5 % to 7 % of this plutonium depending upon the expected burn-up of this new fuel. In turn the MOX spent fuel still contains 3 to 4 % of plutonium, with a fissile content around 60 %. This MOX fuel can then be reprocessed and its plutonium recycled again.

When both uranium and plutonium recovered from spent fuel through reprocessing, are recycled in the LWRs, the fuel cycle is "closed", as all valuable materials are used for energy production, and only the ultimate waste is disposed of.

Most of the major countries engaged in a long term development of their nuclear energy program in Europe and in Japan have chosen to close and improve their fuel cycle by reprocessing and recycling.

Another "source" of plutonium will arise from the inventories of in-excess weapons plutonium resulting from the dismantlement of nuclear weapons in the USA and in Russia. According to figures made available to the public, each country will have to manage around 50 metric tons of such excess weapons plutonium.

Economics and Politics of Energy
Edited by Kursunoglu *et al.*, Plenum Press, New York, 1996

A major, reference Study: "Management and disposition of excess weapons plutonium", has been recently published by the US National Academy of Science (NAS), recommending to study further two main options for such plutonium disposition: vitrification and final disposal, or fabrication of MOX fuel and utilization in reactor.

This paper will review the status of the civilian MOX fuel cycle in France and Europe and the coherence of the programs, then the paper will examine the advantages of MOX fuel utilization in LWRs to get rid of the weapons plutonium.

THE MOX FUEL UTILIZATION IN FRANCE AND IN EUROPE

- It has been eight years since the first MOX fuel assemblies were loaded in France in the EDF St Laurent B1 reactor, a 900 MWe PWR.
  EDF is now fully engaged in a large-scale plutonium recycling program with MOX fuel to be loaded in its existing 900 MWe reactors:

  * Today, seven reactors are already loaded and operate with 30 % MOX fuel and 70 % uranium fuel.
  * 9 additional reactors are fully licensed to receive MOX fuel.
  * 12 additional reactors are technically capable to operate with MOX fuel. However they require to follow the administrative procedure for licensing, including a public enquiry.

- In the rest of Europe, today 13 reactors have been loaded or are in the process of being loaded with MOX fuel assemblies in Belgium, Germany and Switzerland and 9 more reactors are licensed and ready for MOX fuel loading.
- So the European status of MOX fuel use in reactors is as follows: 20 reactors are "moxified", out of 38 fully licensed.
  And around the year 2000, up to 50 reactors will be operated with a percentage of some 30 % of their core with MOX fuel.
- The in-core experience as reported by the European utilities proves no operational difference between UO_2 fuel and MOX fuel in terms of performance and safety : a few examples are given below:

  * In France EDF has now already accumulated more than 20 reactors x years of experience and finds no difference in the fuel assemblies in-core behaviour. More, the load following modes for "Moxed" reactors (frequency adjustment and power level modulation) have been successfully tested on the St Laurent B reactor and a generic licensing authorization has been granted by the French Safety Authorities.
  * In Switzerland, the two Beznau reactors are operated with MOX fuel since 17 years and high burn-ups in excess of 45,000 MWd/ton have been achieved without difficulties.
  * In Germany, since MOX recycling was started in 1972 at Obrigheim (KWO) a total of over 350 MOX fuel assemblies for PWRs and BWRs have been loaded, half of which having reached burn-ups above 40,000 MWd/ton.
  * In Belgium where the first MOX fuel irradiation experience took place in the BR3 reactor in 1963, commercial MOX fuel loading has started this year in two reactors (Tihange and Doel).

THE COMMERCIAL RECYCLING INDUSTRY AND THE COHERENCE OF THE EUROPEAN STRATEGY

In Europe the industry of the back-end of the fuel cycle: reprocessing and MOX fuel fabrication, has reached its maturity.

COGEMA plays a leading role in these sectors, by offering to power utilities a coherent industrial system of facilities:

- La Hague UP3 reprocessing plant, started in 1990: 800 tons of spent fuel per year.
- Cadarache MOX fuel fabrication facility, started in 1992: 30 tons of MOX fuel per year.
- La Hague UP2-800 reprocessing plant, started in 1994: 800 tons of spent fuel per year.
- Marcoule MELOX MOX fuel fabrication facility, started in 1995: installed capacity 120 tons of MOX fuel per year, to be increased to more than 160 tons per year (PWR and BWR fuel) in a few years from now.

Furthermore, through its partnership with Belgonucléaire, COGEMA is managing and marketing the capacity of the Dessel MOX fabrication facility (35 tons of MOX fuel per year).

For the year 2000, two new MOX fuel fabrication plants are now under project, one in Great Britain (Sellafield) and the other in France (La Hague), bringing the total MOX fabrication capacity in Europe to some 400 tons per year.

So by the end of the century, the quantities of plutonium separated by reprocessing (2,500 tons of spent fuel), the quantities of plutonium introduced in MOX fuel fabrication (400 tons of MOX fuel per year) and the quantities of plutonium recycled as MOX fuel in European reactors (around 50) will be balanced: 25 tons of plutonium per year, representing an annual energy production equivalence of more than 25 millions T.O.E.

So the global inventory of European plutonium is being well mastered, underlining the overall strategy coherence for reactor-grade plutonium management.

THE RATIONALE OF WEAPONS PLUTONIUM RECYCLING AS MOX FUEL IN LWRS

The industrial experience in the civilian MOX fuel cycle, readily available from the utilities and the fabrication capabilities, give a strong reference for MOX fuel fabrication from weapons plutonium, a major choice for the current issue of weapons plutonium disposition in Russia and the United States. It compares favorably in all respects to the other main alternative being considered, vitrification of weapons plutonium in a matrix containing fission products.

- Technically, MOX fabrication of weapons plutonium will not present serious difficulties:
  In terms of thermal power of the material in process, health physics and safety (α, γ and neutrons radioactivity is notably lower), the properties of weapons plutonium make it easier to handle than reactor plutonium. The only constraint would come from criticality, leading to the need for some specific adaptation of the civilian MOX technology, for example smaller size equipment in the first part of the facility. But this is not perceived as a difficulty : in fact the specific adaptation required would depend on the form under which weapons plutonium would be made available to the

commercial cycle. By comparison, plutonium vitrification was never studied seriously and may imply significant R and D costs and delays.

- Industrially, the disposition of a 50 tons weapons plutonium inventory through moxification would represent an important but manageable program, using existing power reactors.

  For instance and for various realistic scenarios:

  * A MOX fuel fabrication capacity from 80 tons per year to 170 tons per year (processing from 3 tons of Pu to 7 tons of Pu per year). This is in the range of the civilian MOX fuel fabrication plants capacities.
  * A number of reactors ranging from 12 to 20 (with current LWRs loaded with 30 % MOX). An alternate scenario would require two large dedicated 100 % MOX new reactors (EPR type).
  * An overall time scale depending primarily on licensing procedures, lasting from 12 years to 15 years in case of use of current reactors and 20 years or more in case of dedicated reactors to build.

- From the natural resources conservation point of view, the electricity production of a 50 tons inventory with current MOX fuel in-core performances will attain more than 350 TWh, or the equivalent of 50 million T.O.Es.

  Vitrification as it considers plutonium as a waste, will not take any energy advantage of the inventory.

- Economically, the use of weapons plutonium in MOX fuel for LWRs will offer a better performance than vitrification of this plutonium.

  According to our evaluation, within a range of reasonable cost assumptions for uranium, enrichment and fabrication, the cost of a kg of fabricated weapons plutonium MOX fuel is cheaper or comparable to a kg of the usual UO_2 fuel with the same in-core (burn-up) performance. The competitive advantage of MOX fuel increases with the burn-up level and with fuel standardization. This may explain the potential interest of some utilities for burning such MOX fuel in their reactors.

  Vitrification of weapons plutonium on the other hand has only a cost without any return.

- Regarding the non-proliferation aspects, using weapons plutonium in LWRs offers two distinct intrinsic advantages over vitrification :

  ◊ Actual reduction of the quantities of plutonium: the material fissioned in the reactor to produce electricity is actually burned.

  ◊ Significant lowering of the isotopic grade of the plutonium through in-core irradiation, resulting in unpractical quality for weapons use.

In the case of vitrification, the quantity and weapon-grade quality of plutonium remain unchanged, and a potential threat after decay of the fission products (a few tens of years).

CONCLUSION

The civilian plutonium recycling industry has accumulated an operational experience with MOX fuel which will prove helpful for a sensible solution to the weapons plutonium disposition issue.

Such a solution has been identified as a major strategy in the conclusions of several recent committee reports in Europe and the USA analyzing the weapons plutonium disposition program.

In its Report, the National Academy of Science panel "judges that with a prompt decision to proceed in this direction -and given high national priority assigned to the task-fabrication of (weapons plutonium) W-Pu-MOX fuel could begin in the United States as soon as the year 2001".

The ANS "Blue-Ribbon Panel" concluded in a Report released a few weeks ago that "the use of plutonium as fuel in current commercial reactors is already taking place routinely in several countries and can, therefore, be implemented with little delay".

And a few days ago at the Winter American Nuclear Society Conference in San Francisco, Nobel Laureate Glenn Seaborg in a major presentation supported strongly the irradiation of excess weapons plutonium in MOX fuel.

In addition, the Russian authorities have clearly stated their own determination to recycle their excess weapons plutonium and have already started preparations to that effect, with the technical support of COGEMA and the French Atomic Energy Commission.

The European industry stands ready to contribute to such weapons plutonium disposition programs in the US, through Lead Test Assembly programs, transfer of technology and know-how, technical assistance and engineering services, and "quick-start" MOX fabrication programs to be considered.

NUCLEAR MATERIALS SAFEGUARDS FOR THE FUTURE*

J. W. Tape

Los Alamos National Laboratory
Nonproliferation and International Security Division
Los Alamos, NM 87545

ABSTRACT

Basic concepts of domestic and international safeguards are described, with an emphasis on safeguards systems for the fuel cycles of commercial power reactors. Future trends in institutional and technical measures for nuclear materials safeguards are outlined. The conclusion is that continued developments in safeguards approaches and technology, coupled with institutional measures that facilitate the global management and protection of nuclear materials, are up to the challenge of safeguarding the growing inventories of nuclear materials in commercial fuel cycles in technologically advanced States with stable governments that have signed the nonproliferation treaty. These same approaches also show promise for facilitating international inspection of excess weapons materials and verifying a fissile materials production cutoff convention.

INTRODUCTION

The management, protection, and control of nuclear materials to prevent their unauthorized use (safeguards) is one of a few remaining technical barriers to the proliferation of nuclear weapons. National (also called domestic) safeguards and security are designed to protect nuclear materials from misuse, including diversion and theft by adversaries of the State operating at a subnational level but possibly including terrorism sponsored by an external state. Domestic safeguards and security systems are the first line of defense in controlling and protecting nuclear materials and are a fundamental building block of international arrangements for managing and controlling nuclear materials. International safeguards are designed to verify that States are meeting their nonproliferation commitments to place nuclear materials and facilities under inspection through bilateral, regional, or multilateral arrangements.

---

* Portions of this paper are drawn from the report of the Subpanel on Safeguards and Security found in the American Nuclear Society Special Panel Report "Protection and Management of Plutonium," (August 1995).

Economics and Politics of Energy
Edited by Kursunoglu *et al.*, Plenum Press, New York, 1996

The nuclear materials of interest to proliferation can be divided into three categories as shown in Fig. 1; materials in the five nuclear weapons states defined by the nuclear nonproliferation treaty (NPT), materials under international safeguards in civilian applications, and materials outside international safeguards. New initiatives to address nuclear materials control issues outside the realm of traditional safeguards include the U.S. offer to place excess weapons materials under some form of international inspection and the proposed fissile materials production cutoff convention, which would ban future production of fissile materials for weapons purposes or outside international safeguards.

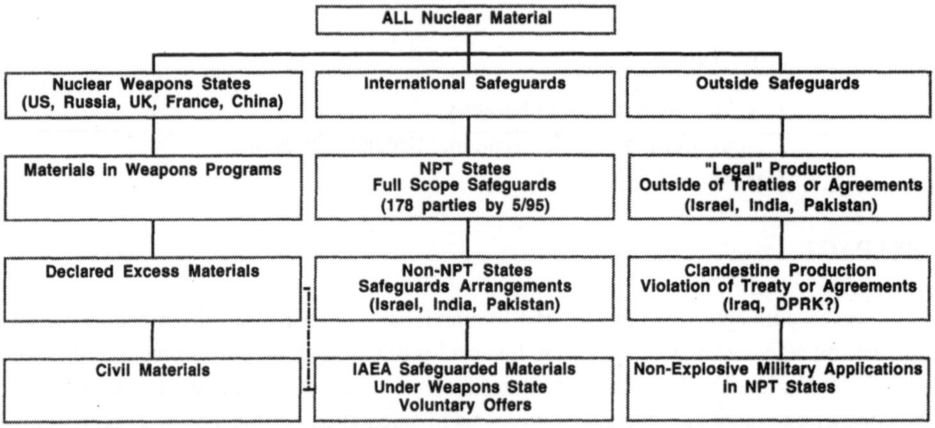

Figure 1. Schematic breakdown of global nuclear materials into three broad categories in nuclear weapons states, under international safeguards, and outside safeguards.

Because of their value and sensitivity, all nuclear materials are likely to be under some form of domestic safeguards to protect against theft and unauthorized use. Domestic safeguards systems are usually described in terms of three elements: physical protection, materials control, and materials accounting. In practice, materials control and accounting activities are often managed as part of an overall materials control and accounting (MC&A) system.

Materials accounting and containment and surveillance (C/S) are fundamental elements of international safeguards; however, international safeguards are being strengthened to include broad measures that go well beyond these two traditional safeguards measures to include improved access to information and sites. The primary goal of the strengthening measures is to improve the International Atomic Energy Agency's (IAEA) ability to detect undeclared nuclear materials or facilities in countries that have signed the NPT. Although advancements in traditional IAEA safeguards are usually evolutionary, some of the new measures to be included in strengthened safeguards are more revolutionary and follow directly from the lessons learned in Iraq following the Gulf War.

International safeguards have no direct equivalent of physical protection (considered the responsibility of the State); however, the IAEA does support the development of common standards for the protection of nuclear materials and the dissemination of knowledge about physical protection measures and systems. The United States government also encourages the transfer of physical protection technology and methods by means of bilateral discussions and technology transfer programs. Furthermore, modern integrated systems that combine features of C/S, process monitoring, and continuous, unattended assay of materials can provide timely warning of anomalous conditions in facilities under international safeguards and thus perform some of the prompt-detection functions of a physical protection system.

Integrated systems are the key to successful safeguards for either domestic or international applications. Defense-in-depth is a fundamental principle of design that is often misunderstood by those who attempt the very difficult task of assessing the effectiveness of safeguards systems. Timely detection of diversion might be, for example, the result of surveillance (by humans or sensors), near-real-time accounting systems, process monitoring, portal monitors, access control systems, or intrusion detectors. Diversion is deterred by all these elements plus an effective accounting system that can provide an audit trail confirming that materials are in fact missing from a facility or process. No one element is the "most important"; they must work together for good materials safeguards.

Safeguards systems designed with a defense-in-depth approach can reduce the risk of diversion or theft from declared fuel cycles to acceptable levels for reasonable costs. Modern domestic safeguards and security systems employed by facilities under regulation by strong, stable governments, when coupled with strong response elements, are commonly viewed by safeguards professionals as providing adequate protection against subnational threats. In this regard, it is important for safeguards systems design to take into account the local differences in threats to nuclear materials and facilities. Concern remains for protecting nuclear materials in unstable parts of the world and protecting against the determined State proliferator.

In considering the role of safeguards in the nonproliferation regime and their relationship to the growth of civilian nuclear power, it is important to recognize that the route to proliferation has been, and is most likely to remain, the use of clandestine, dedicated weapons production facilities either outside any treaty agreements or in violation of agreements. Thus for the nuclear power fuel cycle, the role of domestic safeguards is to maintain sufficient assurance that materials are protected from subnational threats, and the role of international safeguards is to provide assurance that declared materials have not been diverted for nuclear explosives or purposes unknown and that (declared) fuel cycle facilities are not misused for unsafeguarded production of weapons-usable materials. The possibility of clandestine production of nuclear weapons materials has little bearing on safeguards for the nuclear power fuel cycle* except to note that safeguards systems need only be good enough to motivate the determined proliferator to produce materials directly rather than diverting them from safeguarded facilities.

Safeguards must be an integral part of the nonproliferation regime that considers the motives of States or terrorist groups, their capabilities, the number and kinds of nuclear devices they desire, and the global availability of all weapons-usable nuclear materials, not just plutonium. A balanced approach that considers all these factors will provide guidance as to the required performance and cost of safeguards systems. Proposals to eliminate commercial nuclear power, restrict the use of plutonium, and/or to dispose of nuclear materials must be evaluated in this broader context.

TRENDS FOR THE FUTURE

The recent American Nuclear Society special panel report, "Protection and Management of Plutonium," noted that while nuclear power is likely to be an important

* The ANS Special Panel report contains an extensive discussion of this point.

component in meeting future energy demand, it does not follow that there needs to be widespread adoption of plutonium recycling. In particular, it seems that the facilities posing the greatest proliferation risk, reprocessing and uranium enrichment plants, do not need to be located outside advanced industrial states with excellent proliferation credentials. Thus, in considering future requirements for nuclear materials safeguards systems, it is possible to think about three general classes of nuclear power use and infrastructure: states with full fuel cycle facilities with reprocessing, enrichment, and possibly breeder reactors; states with MOX-fueled reactors with the fuel supplied by and spent fuel returned to states with full plutonium recycling facilities; and states with reactors using low-enriched uranium fuels only with safeguarded storage of spent fuel or return of spent fuel to supplier states. Therefore, it is not necessary to imagine the need for the most elaborate systems everywhere, but only in selected locations. And while the amount of separated weapons-usable materials is certainly an important parameter for safeguards, the number of geographical locations is probably a more important challenge to the regime.

Safeguards issues for the future can be divided into political/institutional measures and technical measures; however, the two are linked. Advances in safeguards technology can influence institutional measures and political requirements can demand more of technology.

Institutional/Political Measures--Global Nuclear Materials Management

Global nuclear materials management refers to an integrated, systematic approach to thinking about and dealing with nuclear materials and incorporates ideas, not all of which have equal value, that have been proposed for a number of years and in a variety of forms. Although the proposals are grouped under institutional/political measures, most have a strong technical component as well. The primary reason for their categorization as institutional/political is that the barriers to implementation are not primarily technical.

Minimizing Amounts and Locations of Separated Materials. Although we understand how to effectively safeguard them, the storage of large quantities of direct-use materials for long times at multiple locations should be avoided. Safeguards costs can be reduced by sizing the elements of the fuel cycle to separate only those amounts of attractive materials that are needed, by minimizing the time that those materials are in a direct-use form, and by limiting the number of storage locations. It is important to note, however, that spent fuel still must be safeguarded against diversion by the State (and sabotage by terrorist groups), and that spent fuel becomes more attractive with time as the fission products that provide the radiation barrier decay. Balancing materials supply with demand is made more complex by the economies of scale of building and operating reprocessing plants. Reprocessing plants that are part of the international nuclear fuel cycle, such as those in the UK and France, must also be prepared to deal with the uncertainties of the flows of materials to and from their facilities through the use of safeguarded and secure interim storage arrangements.

Materials from nuclear weapons dismantlement, which are attractive to begin with and exist in large quantities, can be dealt with in a variety of ways including special, highly secure storage; timely use in forms that are more proliferation resistant; disposition through reactor or accelerator burning; mixture with high-level waste; or direct deep burial. The near-term concerns for dealing with materials from weapons dismantlement are first to protect them during dismantlement, storage, and ultimate disposition and then to ensure irreversible arms reduction. The nonproliferation benefits accrue from minimizing the terrorist or insider threat to the materials and for showing more intangible aspects of nonproliferation leadership by demonstrating reduced reliance on nuclear weapons.

International (dual key) Control on Sovereign Territory. Stocks of attractive nuclear materials not needed for immediate processing could be placed under dual access controls of the State and the IAEA. The State would declare its requirements for withdrawals from the store, which would be made under observation by State and IAEA inspectors. The control exercised by the IAEA would not include the right to veto a materials movement, rather it would provide an additional layer of containment and surveillance. Although the materials would be stored on the State's territory and would obviously be under the ultimate control of the State, dual controls would provide an important confidence-building function regarding the State's nonproliferation commitments. Formal mechanisms of this kind have been proposed in the past (for example, international plutonium storage), but never implemented. In practice, with the evolution of continuous surveillance and the possibility of remote monitoring (see Technical Measures section of this paper), the IAEA can have real-time knowledge of materials movements similar to what would be achieved under dual-key control.

Transparency. Transparency measures are not well defined by the international safeguards community but can include more openness about the purposes of nuclear activities, plans, and inventories of a State as well as more access by inspectorates to facilities, declared and nondeclared, in the State. The IAEA's "93+2" program to strengthen safeguards relies on improved access to information and sites that, if granted by the State, will provide considerably increased transparency about the State's nuclear enterprises. Transparency builds confidence and provides information that can be used to more effectively allocate safeguards resources.

Intelligence Sharing. Although the IAEA can go a long way in improving its ability to detect undeclared activities and materials, it cannot hope to acquire the resources for detecting these activities that are used by many nations. Furthermore, the Agency has no authority to find non-compliance in regard to activities of States not covered by safeguards agreements. Careful intelligence sharing among nations and international organizations, including the IAEA, is therefore an essential element of the nonproliferation regime. Integral to this effort is improving the intelligence communities' ability to detect proliferation and in particular, the production, theft, or smuggling of nuclear materials.

Improved States' Systems and Regional Safeguards. Global nuclear materials management begins with the development and implementation of protection and control systems by those with responsibility for the materials: facility operators and the State. Improvements in these systems can make significant contributions to international safeguards in stable regions with cooperative governments. Regional approaches such as EURATOM or ABACC (the joint Brazil/Argentine control commission) also facilitate global controls, serve as important confidence-building measures for nonproliferation, and have economic benefits through the use of shared resources.

Strengthened Physical Protection. Physical protection for nuclear materials must be balanced to meet the local threat on a global basis. The international safeguards community has made major strides in disseminating physical protection standards, but more can and should be done, possibly including a role for the IAEA as an invited (by the State) independent auditor or assessor of physical protection systems. Advanced monitoring technologies can also play a physical protection function by providing timely warning of diversion even in the international safeguards context.

Proliferation-Resistant Measures. Making nuclear materials inherently less useful for rapid or simple fabrication into a nuclear explosive device can be effective in reducing the terrorist or insider threat. It does not, however, significantly reduce the risk of diversion by the State in the international safeguards context, although it may delay the time from diversion to fabrication. Examples of proliferation resistant measures are the integral fast reactor, in which fission products are never completely separated from plutonium during reprocessing that is integral to the reactor; protecting materials through the use of "natural" barriers such as high-radiation fields (for example the National Academy of Science recommendation on the "spent fuel standard,"); and coprocessing of mixed oxide materials.[1] The technical measures are generally well understood; however many policy and political barriers remain.

Supporting the IAEA. The IAEA and the international safeguards community are only a part of the nonproliferation regime; however, the Agency plays a central role in enhancing international security and the security of all nations. As such, the Agency's costs are a bargain. It is important that the IAEA be provided the resources—financial, technical, and personnel—to meet the challenges of a post-cold war world with a growing reliance on nuclear power.

Technical Measures

The continual evolution of improved technology is the key to cost-effective global nuclear materials management and control. In the face of the growing materials inventories and the number and sophistication of nuclear facilities, both domestic and international safeguards for declared facilities must rely on technology to provide continuous near-real-time knowledge of the amounts, locations, and movements of nuclear materials. To provide assurance that these materials are still where they are supposed to be and have not been stolen or unintentionally released to the environment, they must be measured and inventoried, pathways to the environment must be monitored, and records must be maintained. The detection of undeclared materials or facilities by safeguards inspectorates will rely on sensitive detection methods based on environmental sampling near facilities or over wide areas, as well as on other detection and monitoring approaches.

Nuclear materials control technology has been employed since the beginning of the nuclear age. In the past 30 years much of the emphasis has been on developing methods to detect and measure nuclear materials (primarily uranium and plutonium) rapidly and nondestructively, to improve the precision and accuracy of destructive analysis methods, to provide effective physical protection of materials and facilities, and to ensure the continuity of knowledge about nuclear materials. These technologies span a broad range of instrumentation and systems that includes portable equipment carried by inspectors; on-site instrumentation for use by plant operators; continuous, unattended monitoring and assay equipment for on-site inspection; containment and surveillance measures; and remote monitoring concepts for safeguards, environmental, and proliferation detection applications.

Sensors and monitoring devices are useful only if the information acquired by the instrumentation is organized, analyzed, and displayed in a way that can be used by human decision-makers who must provide assurance that nuclear materials have in fact remained under control. Thus, information management technologies are an integral part of nuclear materials management and control systems.

Although existing technology for nuclear materials control can be used to meet the demands of today's problems, new technology is needed to improve the effectiveness and efficiency of both domestic and international safeguards systems. Nuclear technology is not stagnant, and nuclear materials control technologies must keep up. Instrumentation for safeguarding nuclear materials can be made to provide more timely results, to be more portable, to operate continuously in harsh environments, and to be less expensive. Information management, the key to effective nuclear materials control, is based on the revolution in computer and communications technology. Hardware and software developments for improved analysis of materials inventories and transfer information as well as physical protection or containment and surveillance information have the potential to alert nuclear materials managers or inspectors to problems where human analysis would be impossible or not timely.

Trends for future safeguards technical developments can be divided into four categories:

- *sensor improvements* including, for example, enhanced sensitivity methods for environmental monitoring and measurement methods with improved accuracy for materials accounting;
- *information management advances* such as detecting anomalies in large databases, computerized materials control and accounting systems, sensor fusion, and remote transmission of large quantities of data;
- *integrated, automated, facility-level safeguards systems* that provide defense-in-depth such as those currently being demonstrated in the early implementation phase by EURATOM inspectors and plant operators; and
- *safeguards effectiveness evaluation methods* to provide confidence to safeguards authorities and the public that nuclear materials are in fact under control.

The following sections describe briefly some of the current developments and future trends for safeguards technology; however, this should not be taken to be a comprehensive review of the subject.

Sensors. Needed advances in sensor technology range from improvements in seals to make them more secure, reusable, and remotely readable to the development of nondestructive assay (NDA) methods for the quantitative assay of nuclear materials with improved precision and accuracy that can operate continuously and unattended. Environmental monitoring methods employed to detect undeclared facilities or production also represent advances in sensors for safeguards.

Safeguards applications of seals provide continuity of knowledge about nuclear materials by indicating whether or not an item, container, or portion of a facility has been accessed between inspections or authorized accesses. Simple mechanical seals can only indicate that the seal has been broken (and that access occurred at some time by someone). "Smart" seals have the potential to indicate access times and dates over multiple periods, to identify the user, and to be verifiable in situ or possibly remotely.[2,3] When coupled with other information, such as video surveillance, it is possible for safeguards authorities to determine authorized access to materials and locations and to detect unauthorized activities. This technology permits authorized plant operations to continue without the continued replacement of seals and improves the overall knowledge of operations. Smart seals can also be employed with automated, unattended material measurement devices to link assay results with item identity without inspectors being present.

Physical protection and material control systems rely on sensors to indicate unauthorized or otherwise anomalous activities in nuclear facilities by providing alarms in real time. These types of sensors include intrusion detection systems that are typically deployed on facility perimeters and item and personnel monitoring and tracking systems. Technology development thrusts for these sensors include improving sensitivity and specificity, reducing false alarm rates, reducing costs, authenticating data, and improving data transmission characteristics for collection in central alarm stations. Deployment of these sensors can reduce materials inventory frequencies by providing real-time assurance of item integrity and can reduce the threat of unauthorized activities by facility personnel (the "insider threat").[4]

Surveillance systems based on advances in video and computer technology are being made more reliable and can play a role in real-time monitoring of nuclear facilities.[5] As with many of the other developments in sensor technology, one goal is to ensure the continuing security of nuclear materials or storage locations and thereby reduce inventory frequencies with resulting savings in costs and personnel radiation exposures. In the most straightforward applications, images of static scenes are digitized and compared with the same scene at later times to detect unauthorized access to specific items. By differencing the digitized before and after images, it is possible to detect even small disturbances of items that would be undetectable by human observation.[6] (See Fig. 2)

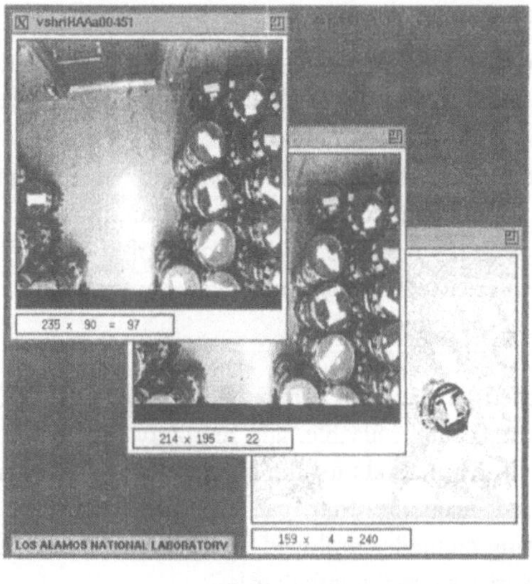

Figure 2. Digitized video images of a nuclear materials storage vault. A camera views storage containers (drums) from overhead. The before, after, and difference images are displayed, indicating that one drum was disturbed during a vault access.

Portal monitors, (Fig. 3) one of the key detection elements in the safeguards systems of nuclear facilities, are designed to detect the unauthorized passage of nuclear materials across specified boundaries. The passage of nuclear materials at facility access points can be monitored with simple hand-held search equipment or fixed detectors designed to monitor personnel or vehicle traffic. Most monitors depend on the detection of gamma radiation to indicate the presence of nuclear materials; however, the addition of neutron detection capability enhances the detection of plutonium. Advances in portal monitor technology include the use of improved detectors, sophisticated electronics and computer software to account for varying backgrounds, and the integration of radiation portals with other portal detection technologies such as metal detectors and access control devices.[7] Through the use of these integrated technologies, portals can be made unattended with remote transmission of alarms.

Figure 3. Vehicle radiation portal monitor. Detectors under the canopy and in the roadbed sense gamma radiation.

Improvements in NDA technology are needed to achieve accuracy requirements for measurements by reducing the sensitivity to bias effects caused by unknown sample characteristics. Gamma-ray tomography is being developed to deal with the assay biases that result from inhomogeneous material distributions in containers of nuclear materials and wastes. Although it is still early in the development of this method, tomographic gamma scanning (TGS) has demonstrated the capability to assay 55-gallon drums of heterogeneous materials containing plutonium with accuracies that greatly exceed those achieved in conventional gamma ray scanning.[8] (See Fig. 4)

Although tomography is not recommended for all gamma-ray assay applications, it provides the only solution for difficult assay situations, short of unpacking the entire container and conducting nondestructive or destructive analyses on the contents: a very expensive and time-consuming proposition.

Neutron multiplicity analysis provides a similar advance for neutron-based nondestructive assays that can be biased by unknown matrix and multiplication effects. The development of high-efficiency neutron coincidence counters with multiplicity electronics to measure single, double, and triple neutron events (coincidences) provides information that permits the determination of sample mass, neutron multiplication, and (alpha, n) rate.[9] Multiplicity counters provide improved accuracy assays and require fewer reference materials to calibrate (Fig. 5).

Continuous, unattended monitoring and assay systems have the potential to improve nuclear materials management by providing timely, low-cost information on amounts and locations of nuclear materials in bulk-handling facilities. For modern,

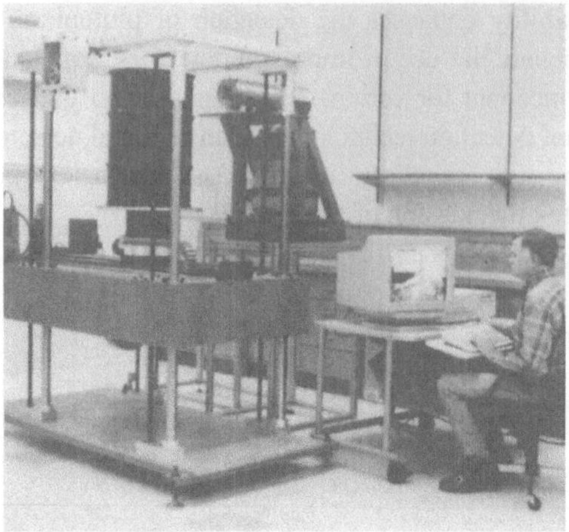

Figure 4. Prototype Tomographic Gamma Scanner designed to nondestructively determine plutonium content of large containers.

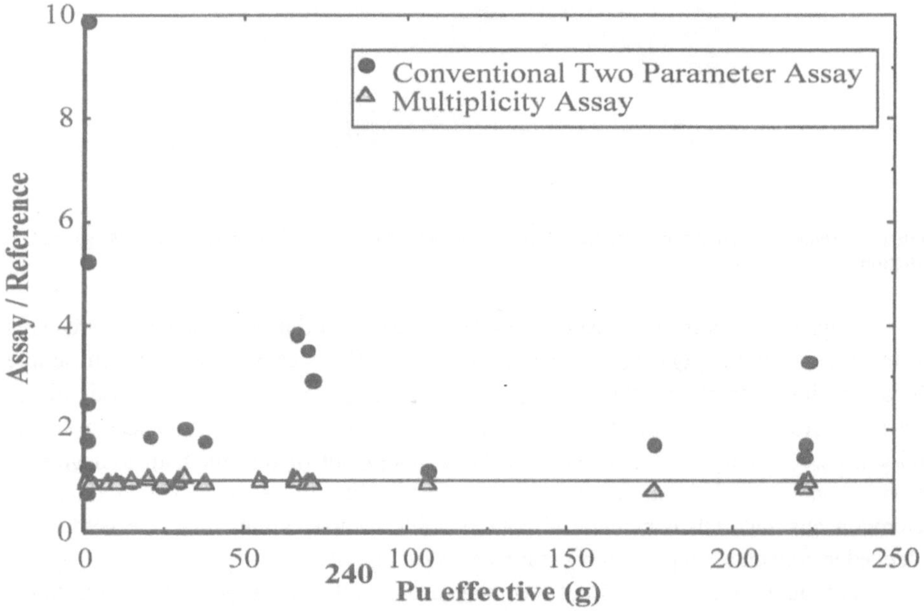

Figure 5. Ratio of nondestructive assay value to the known reference value plotted vs. the effective ^{240}Pu content of the sample. Multiplicity assay provides significant reduction in assay bias over conventional two-parameter neutron coincidence counting.

automated nuclear facilities, such instrumentation is the only possible solution to tracking and accounting for nuclear materials throughout the process. The first significant attempt to employ continuous, unattended assay and monitoring equipment was the project undertaken for the Japanese Power Reactor and Nuclear Fuel Corporation (PNC) Plutonium Fuel Production Facility (PFPF) (Fig. 6) beginning in 1988 to provide a suite of instruments to be installed in the plant and interfaced with the robotic fuel fabrication equipment to provide safeguards information for the IAEA.[10]

Figure 6. Nondestructive assay equipment installed at the final fuel assembly transfer point in the Japanese Plutonium Fuel Production Facility. Fuel assemblies are moved and plutonium content of assemblies is determined automatically without inspector presence.

This project has demonstrated the value of using continuous, unattended equipment by improving the knowledge of materials and reducing costs for both the operator and the inspectorate. Reliable communications, hardware, and software are essential for these applications and can be achieved through careful systems engineering, including redundancy. Data authentication is also a key consideration for unattended safeguards systems, and can be achieved through a number of means including encryption and the use of integrated surveillance systems.

Environmental monitoring has only recently been considered for application to international safeguards to detect undeclared materials and activities at declared sites or possibly to detect undeclared facilities. In this context, the monitoring activities include sample collection, packaging, transportation, and analysis using a suite of analytical methods that are, for the most part, well established. Samples include cloth swipes, biota, soil, and water obtained at or near nuclear facilities or suspect sites.[11] In another approach that is based on in-situ detection of alpha particle radiation over long distances, it is possible to detect in real time radioactive materials contamination at very low levels on surfaces in a manner that could be used to detect undeclared activities, materials or facilities.[12] Environmental monitoring shows great promise for safeguards; however, considerable work is required to incorporate this "sensor" into routine inspection use, and to improve data analysis and interpretation for unambiguous safeguards conclusions.

Information Management. The management of information is perhaps the most significant area for advancing safeguards technology. The data provided by the sensors described briefly in the previous section is not very useful unless it can be used to make decisions about the status of nuclear materials and the safeguards system. By taking advantage of new computer and communications technologies, safeguards systems can be made to rely less on humans performing mundane tasks, freeing plant operators as well as inspectors and other safeguards personnel for higher-level tasks requiring judgments regarding the performance of the safeguards systems.

Remote monitoring, that is, the transmission of safeguards sensor information to locations removed from the sensors, is a key element for future improvements in the effectiveness and efficiency of safeguards systems, in particular, international safeguards systems. Having large numbers of inspectors on-site is expensive for both the plant operator and the inspectorate. Even when it is desirable to have inspectors on-site, remote monitoring concepts can reduce the numbers required and improve the productivity of the inspectors. Remote monitoring concepts are being developed, field-tested, and evaluated to examine and define the technical parameters of remote monitoring systems, to demonstrate the technical feasibility and political acceptability of remote monitoring, to gain acceptance for the concept in the safeguards community, and to identify legal and institutional constraints in the implementation of remote monitoring.[13] Based on an on-site integrated monitoring system (IMS), which brings a suite of local sensor data into a common database though the use of a local network and provides for some local data analysis for data reduction, the remote monitoring system permits the transmission of information, including video surveillance images, to other locations. Information about nuclear facilities and the status of safeguards at those facilities has long been considered sensitive by the host nations and not to be transferred in large quantities over communications networks. However, improvements in technology, such as public key encryption as well as changing political commitments to transparency in nuclear activities, are expected to enable the use of remote monitoring in many applications in the near future. Remote monitoring technologies can also be employed within the nation by inspectorates and reduce costs.

Normally, containment and surveillance sensors, such as radiation monitors and video camera systems, operate independently, providing information to safeguards authorities through separate channels for later analysis by human operators. On-line integration of video digital data, radiation monitoring, and other sensor data promises to increase sensitivity to detecting events of safeguards interest while freeing inspectors or other safeguards authorities from tedious and time-consuming analysis of large quantities of information. Automated pattern-recognition analysis of these integrated safeguards data sets using neural nets and other methods increases evaluation effectiveness by showing trends, discovering anomalies, and highlighting specific activities for detailed review by inspectors. A fundamental problem is the integration of disparate data, for example, from radiation and video sensors. The video data used to monitor movements are spatial in nature; whereas, the NDA sensors provide radiation levels as a function of time. However, changes in video scenes represent movement and can be quantified using a metric that provides motion levels versus time in addition to the spatially oriented video data. In this approach, the video component becomes another detection element as well as a traditional recording instrument.[14] Another approach to sensor fusion that is being applied to exterior intrusion detection systems is the use of "fuzzy" sets and "genetic" algorithms to combine information from three sensors to increase the reliability of and confidence in distinguishing between real and false alarms.[15] Future developments in combining signals and information from disparate sources in real time will be crucial to the effective operation of integrated safeguards systems.

Modern nuclear materials storage facilities will incorporate elaborate sensor suites to monitor conditions of importance to the environment and to the safety and health of the workers. In addition, both domestic and international safeguards systems could use this sensor information. Using common sensors and databases, a single integrated system can provide information to multiple users, thus reducing costs and intrusions on facility operations. The primary technical challenge is providing the right information to the right user, who may be at a remote location, while protecting any sensitive information from unauthorized disclosure. Applications of common communications systems, such as the Internet, and the use of information security systems, are being demonstrated and show promise for the use of these technologies in the future.[16]

Local area networks can provide the basis for a computerized near-real-time material accounting system that captures the accounting data as close to the source of that data as possible.[17] (Fig. 7) Process operators can enter the data directly into the system immediately after each reportable occurrence and process control systems can also provide data in real time. The database of material information is updated as soon as the information is entered. Implemented using a client/server approach, these systems take advantage of the reduction in computer hardware costs by using a powerful server to maintain the material database and personal computer "clients" to provide the user interface.

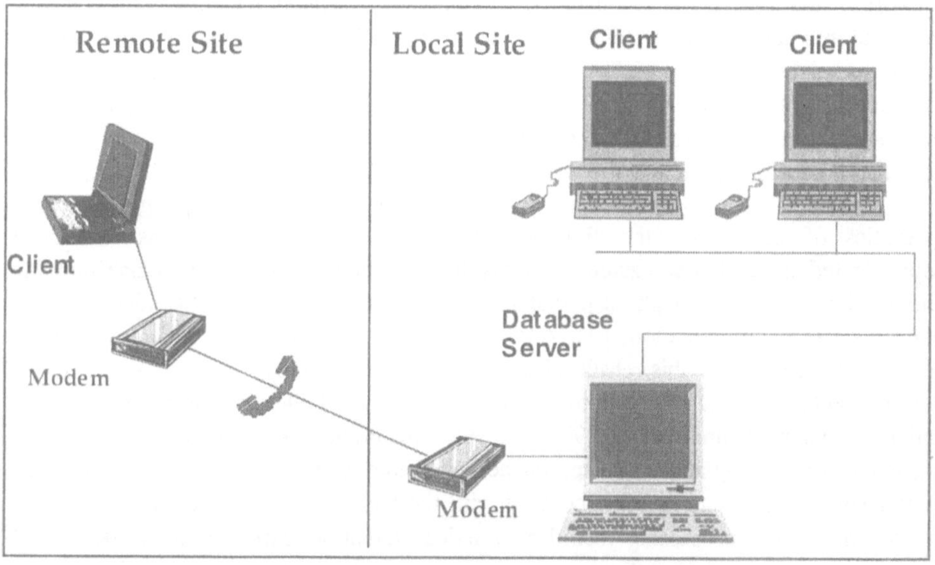

Figure 7. Conceptual Local Area Network Material Accounting System.

Advances in sensors and databases permit the accumulation of vast quantities of information about nuclear materials and the operation of nuclear facilities. The majority of this information is indicative of normal operations of no safeguards concern. The challenge for the safeguards systems is to find information that may indicate an attempted diversion or theft of nuclear materials buried within a mountain of other data. The first step in analyzing safeguards data is to identify anomalies: something that is not normal.

diversion or theft of nuclear materials buried within a mountain of other data. The first step in analyzing safeguards data is to identify anomalies: something that is not normal. Within the set of anomalous events lie the indicators of events of safeguards significance. Anomaly detection schemes can be applied to inventory difference data from materials accounting to identify losses of materials, to sensor data to improve alarm detection performance, and to large databases of materials transaction information to indicate process upsets or diversions. Approaches to anomaly detection include classical statistical methods,[18] the use of fuzzy logic,[19] neural networks,[20] and rule-based expert systems.[21] Process modeling and simulation can also be used to predict future behavior of systems and to detect deviations from the expected condition of processes.[22] Advances in computing and similar information management requirements in many other fields are expected to make significant contributions to the detection of anomalies in safeguards information systems.

Facility Monitoring and Safeguards Systems. The ability to safeguard nuclear materials in large bulk-handling facilities such as mixed oxide (MOX) fuel fabrication plants and reprocessing plants of the kind being operated in Europe and under construction in Japan depends on developing and implementing systems that go far beyond the materials accounting and containment/surveillance measures that were adequate for the facilities of the early commercial nuclear fuel cycle. To provide confidence in safeguards for these new plants, it is necessary to employ a defense-in-depth approach that goes well beyond a periodic determination of material unaccounted for (MUF) and a review of C/S data. Large numbers of sensors integrated into real-time information management systems must be employed and operated by the facility operators under the supervision of national or regional authorities and under the inspection of international agencies. The systems must be designed into the plants with the cooperation of the operator and must provide a degree of redundancy and overlap such that no one component is the weak link in the system.

Faced with the operation of a number of large scale facilities processing significant quantities of plutonium, the EURATOM safeguards inspectorate has developed sophisticated integrated safeguards systems for a number of plants in France and the United Kingdom.[23] Although each system is designed around its facility and employs plant-specific instrumentation, EURATOM and the plant operators have developed a common approach to this challenging problem that employs many of the latest developments in safeguards equipment and concepts and builds on safeguards research and development advances of the 1970s and 1980s including, for example, the continuous, unattended monitoring and assay approaches pioneered in the PFPF in Japan mentioned previously. These systems must contend with the high degree of automation employed in modern nuclear processing facilities to reduce radiation exposure to workers and inspectors; however, automation presents an opportunity if the safeguards system can utilize information provided by the plant's process-monitoring equipment.

The approach to safeguarding these plants is to make use of the large quantities of information available about the flows and inventories of nuclear materials from both independent measurement and detector systems, as well as plant equipment, by employing authentication and verification measures at various levels of independence. Also key to the success of safeguards for large-scale plutonium processing facilities is the use of inspectorate-staffed on-site laboratory facilities to provide timely analysis results for samples while avoiding the costly and time-consuming transport of plutonium samples to central laboratories.

For the thermal oxide reprocessing plant (THORP) operated by British Nuclear Fuel Ltd (BNFL), the enhanced approach is "based on

- A high degree of transparency of operations by the provision by the operator in near real time of a well chosen operating data set, verified by EURATOM according to a hierarchy of control levels;
- Comprehensive (i.e., 100%) quantitative flow verification techniques at key points;
- Application of integrated monitoring and containment/surveillance (C/S) techniques for additional internal flow verifications and maintenance of continuity of knowledge;
- Continuous presence of highly trained inspectors fully familiar with the design and operation of the plant as well as the safeguards measures employed."[24]

This approach provides a higher degree of assurance about the nuclear materials and the absence of diversion than could ever be generated by the materials accounting data and statistical analysis alone. It is the operation of these systems and their promise for other new facilities that give safeguards professionals the confidence that safeguards technology and systems implementation can meet the demands of the growing nuclear fuel cycle, including plutonium recycling.

Safeguards Effectiveness Evaluation Methods. As noted in the previous section, modern safeguards systems for large bulk-handling facilities employ a defense-in-depth approach that goes far beyond the use of materials accounting data alone or the examination of C/S records to determine if materials have been diverted. It is important, especially in the context of international safeguards, for safeguards authorities to be able to assess the overall effectiveness of these integrated safeguards systems to assure the public and governments that materials are under control and that States are in compliance with their safeguards commitments. Performance measures based on a statistical analysis of materials accounting data are attractive because they are quantitative and require no independent judgment that could be politically motivated. However, these straightforward measures, even when coupled with assessments of inspection effort and C/S data, do not seem adequate to determine in a uniform fashion that safeguards systems are performing up to standards. This problem, how to combine information from multiple sources to reach conclusions, is of course, related to the problems described in the information management section of this paper regarding combining data from multiple sensors. In this case; however, the need is to examine the entire safeguards system and the validity of the conclusions based on the system. The safeguards community has not accepted any single approach to assessing the performance of safeguards systems, and "expert" judgment still plays a large role. Although the expert's opinions will always be important, further research and application of theories of information combination appear to be warranted for assessment of safeguards systems.[25]

CONCLUSIONS

Safeguards technology development, supported by research and development programs in the U.S.* and overseas, continues to improve the effectiveness and efficiency

* In the U.S. the Department of Energy Office of Nonproliferation and National Security is a major supporter of safeguards research, development, and implementation. The Department of State provides support for implementation of IAEA safeguards.

materials in safeguarded facilities to provide assurance that materials from the power reactor fuel cycle are not being diverted to military explosives. Defense-in-depth approaches that use the latest in sensor and information management technologies can deal with the high-throughput fuel cycle facilities, as demonstrated in facilities in the UK and France. The detection of undeclared facilities, although of great importance to the nonproliferation regime and international safeguards, is not a significant consideration for safeguards on declared facilities and is being improved through the use of, for example, environmental monitoring. With continued technology and institutional development as well as measured growth of the nuclear fuel cycle in a responsible fashion to limited regions, nuclear materials safeguards can meet the challenges of the future. These technologies and institutions also have the potential to facilitate other nuclear materials control regimes, including the international inspection of excess weapons materials in the nuclear weapons states and the verification of a cutoff convention for fissile materials production.

ACKNOWLEDGMENTS

I would like to thank Joe Pilat, Don Close, Tom Burr, Rich Strittmatter, Avigdor Gavron, Joe Claborn, JoAnn Howell, and Debbie Rutherford, for their assistance with this paper; Paul Henriksen for editing the manuscript; and Charlene McHale for helping to keep the paper preparations on track and Paul Henriksen and Charlene McHale for their excellent word-processing skills. As author, I remain solely responsible for its content. The views expressed are my own, and not necessarily those of the University of California, the U.S. Government, or any other institution.

REFERENCES

1. Proliferation resistance criteria and the relationship to the "standards" outlined in the National Academy report are described in detail by D. A. Rutherford, B. L. Feary, J. T. Markin, D. A. Close, K. M. Tolk, D. Mangan, C. Jaeger, R. Moya, R. Duggan, and L. More in "Proliferation Resistance Criteria for Fissile Material Disposition Issues," *Nuclear Materials Management* **XXIV**, 400-406 (July 1995).
2. S. Kadner and K. Ferguson, "The Partnership Approach--New Safeguards Directions," *Nuclear Materials Management* **XXIII,** 1146-1149 (July 1994).
3. P. Chare, P. Detourbet, and W. Kloeckner, "Euratom Experience in Electronic Seal Development," presented at the American Nuclear Society 5th International Conference on Facility Operations-Safeguards Interface, September 24-29, 1995 at Jackson Hole, Wyoming (to be published).
4. D. A. Anspach, I. Waddoups, and E. T. Fox, "Incorporation of Item/Material Attribute System into PAMTRAK," *Nuclear Materials Management* **XXIII,** 545-552 (July 1994).
5. P. Chare, F. Basile, J. Goerten, B. Jargeac, G. Reis, R. Vandaek, H. G. Wagner, "EURATOM Experience with Digital Video Systems," presented at the American Nuclear Society 5th International Conference on Facility Operations—Safeguards Interface, September 24-29, 1995 at Jackson Hole, Wyoming (to be published).

Operations—Safeguards Interface, September 24-29, 1995 at Jackson Hole, Wyoming (to be published).

6. C. A. Rodriguez, "Life and Times: The Development of a Digital Video Surveillance System," *Nuclear Materials Management* **XXIII,** 273-276 (July 1994).
7. R. L. York, P. E. Fehlau, and D. A. Close, "Exporting Automatic Vehicle SNM Monitoring Technology," presented at the American Nuclear Society 5th International Conference on Facility Operations-Safeguards Interface, September 24-29, 1995 at Jackson Hole, Wyoming (to be published).
8. T. H. Prettyman, S. E Betts, D. P. Taggart, R. J. Estep, N. J. Nicholas, M. C. Lucas, and R. A. Harlan, "Field Experience with a Mobile Tomographic Nondestructive Assay System," Proceedings 4th Nondestructive Assay and Nondestructive Examination Waste Characterization Conference, CONF-951091, Idaho National Engineering Laboratory, Idaho State University, 109-137 (October 1995).
9. N. Ensslin, M. S. Krick, and H. O. Menlove, "Expected Precision of Neutron Multiplicity Measurements of Waste Drums, *Nuclear Materials Management* **XXIV**, 1117-1124 (July 1995).
10. H. O. Menlove, R. H. Augustson, R. Abedin-Zadeh, B. Hassan, S. Napoli, T. Ohtani, M. Seya, and S. Takahashi, "Remote-Controlled NDA Systems for Feed and Product Storage at an Automated MOX Facility," *Nuclear Materials Management* **XVII**, 267-273 (July 1989); M. C. Miller, H. O. Menlove, R. H. Augustson, R. Abedin-Zadeh, T. Ohtani, M. Seya, and S. Takahashi, "Remote-Controlled NDA Systems for Process Areas in a MOX Facility," *Nuclear Materials Management* **XVII**, 274-280 (July 1989); S. F. Klosterbuer, E. A. Kern, J. A. Painter, and S. Takahashi, "Unattended Mode Operation of Specialized NDA Systems," *Nuclear Materials Management* **XVII**, 262-266 (July 1989).
11. J. H. Cappis, D. J. Rokop, D. W. Efurd, F. R. Roensch, C. M. Miller, T. Benjamin, and R. E. Perrin, "Actinide Determination and Analytical Support for Evaluation of Environmental Samples for the IAEA," 17th Annual Symposium on Safeguards and Nuclear Materials Management, ESARDA 27, 403-408 (1995).
12. J. E. Koster, J. P. Johnson, and P. Steadman, "Nonproliferation and Safeguarding via Ionization Detection," 17th Annual Symposium on Safeguards and Nuclear Materials Management, ESARDA 27, 443-449 (1995).
13. C. S. Sonnier, C. S. Johnson, S. F. Moreno, E. D'Amato, A. Bonino, J. Bardsley, D. Sorokowski, K. Veveers, M. Cuypers, F. Sorel, B. Richter, G. Stein, K. Koyama, P. Ek, and G. af Ekenstam, "The International Remote Monitoring Project--An Update," 17th Annual Symposium on Safeguards and Nuclear Materials Management, ESARDA 27, 101-106 (1995).
14. J. A. Howell, H. O. Menlove, C. A. Rodriguez, D. Beddingfield, A. Vasil, "Analysis of Integrated Video and Radiation Data," *Nuclear Materials Management* **XXIV**, 162-167 (July 1995).
15. D. S. Fitzgerald and D. G. Adams, "Adaptive Sensor Fusion Using Genetic Algorithms," *Nuclear Materials Management* **XXIII,** 346-351 (July 1994).
16. C. Nilsen and D. Mangan, "Straight-Line: A Nuclear Material Storage Information Management System," *Nuclear Materials Management* **XXIV**, 894-899 (July 1995).

17. J. Claborn and A. Alvarado, "LANMAS Core: Update and Current Directions," *Nuclear Materials Management* **XXIII,** 313-317 (July 1994).
18. T. L. Burr, "Predicting Linear and Nonlinear Time Series with Applications in Nuclear Safeguards and Nonproliferation," Los Alamos National Laboratory report LA-12766-MS (April 1994).
19. A. Zardecki, "Fuzzy Controllers in Nuclear Material Accounting," *Fuzzy Sets and Systems* **74**, 73-79 (1995).
20. J. A. Howell, H. O. Menlove, G. W. Eccleston, R. Whiteson, C. A. Rodriguez, J. K. Halbig, S. F. Klosterbuer, and M. F. Mullen, "Safeguards Applications of Pattern Recognition and Neural Networks," IAEA Symposium on International Safeguards, IAEA-SM-333/112, Vienna, Austria, March 14-18, 1994.
21. R. Whiteson, L. Spanks, T. Yarbro, F. Kelso, J. Zirkle, and C. Baumgart, "Anomaly Detection Applied to a Materials Control and Accounting Database," *Nuclear Materials Management* **XXIV**, 1255-1261 (July 1995).
22. C. A. Coulter, R. Whiteson, and A. Zardecki, "Simulation Study of Near-Real-Time Accounting in a Generic Reprocessing Plant," *Nuclear Materials Management* **XXI**, 486-489 (July 1992).
23. W. Gmelin and W. Kloeckner, "Safeguards in Europe--An Update"; Y. Paternoster, S. Kaiser, P. Chare, Ph. Dossogne, Ph. Beaudoin, P. Molinari, J. Regnier, "Safeguards Activities During the Commissioning Phase of the MELOX MOX Fuel Fabrication Plant"; P. Chare, A. Dutrannois, W. Kloeckner, and M. T. Swinhoe, "Networking of Safeguards Systems"; and J. P. Denkens, H. G. Wagner, G. Landresse, V. Lahogue, and J. Goerten, "An Integrated Safeguards System for Large Scale Reprocessing Plants," all presented at the American Nuclear Society 5th International Conference on Facility Operations-Safeguards Interface, September 24-29, 1995 at Jackson Hole, Wyoming (to be published).
24. S. Kaiser, H. Nackaerts, R. Schenkel, P. J. Chare, H. G. Wagner, R. Howsley, and E. Williams, "Thorp: Authentication, Transparency and Independence," *Proceedings of the IAEA International Safeguards Symposium* **2** (1994).
25. C. Scovel, "The Combination of Information," Los Alamos National Laboratory report LA-UR-95-4408 (1995).

THE U.S.–RUSSIAN HEU AGREEMENT: A MODERN DAY EXAMPLE OF SWORDS INTO PLOWSHARES

Philip G. Sewell

Vice President, Corporate Development
United States Enrichment Corporation
Two Democracy Center
6903 Rockledge Drive
Bethesda, MD 20817

SUMMARY

After four decades and hundreds of billions of dollars spent on the buildup of nuclear weapons, the United States and the Russian Federation have taken a bold step to convert deadly instruments of the Cold War into peaceful and productive resources. The United States and Russia are implementing a historic agreement whereby the United States is purchasing material recovered from dismantled Soviet weapons for use in commercial electricity production. By purchasing Russian highly enriched uranium (HEU), the United States is helping to ensure that it will be used solely for peaceful purposes.

Under the agreement, approximately 500 metric tons of HEU removed from the equivalent of more that 22,000 nuclear warheads from the former Soviet Union will be converted to low enriched uranium (LEU) suitable for commercial power reactor fuel. Once HEU is transformed into LEU, it becomes useless for nuclear weapons. Swords are being transformed into plowshares.

On June 23, 1995, 24 metric tons of low enriched uranium arrived at the United States Enrichment Corporation (USEC) facility in Portsmouth, Ohio. This was the first of many shipments under a $12 billion, 20-year contract between USEC, serving as executive agent for the Unites States government, and Techsnabexport, executive agent for the Ministry of Atomic Energy of the Russian Federation.

In addition to the national security and nonproliferation benefits, the U.S. purchase of Russian HEU is producing a valuable commodity that will be sold in the marketplace enabling some of the costs of disarmament to be recovered. This agreement is also pumping hard currency into a cash-starved Russia that will greatly help its transition to a market-based economy. Russia has committed to use a portion of proceeds from the sale of LEU derived from HEU for conversion of defense enterprises, upgrading the safety of its nuclear power

Economics and Politics of Energy
Edited by Kursunoglu *et al.*, Plenum Press, New York, 1996

plants and environmental cleanup. USEC will gradually phase the material into its supply mix for sale to the enrichment market.

The world's leading supplier of enrichment services, USEC provides uranium enrichment services to more than 60 U.S. and foreign electric utilities, and as such, is uniquely situated to provide a market outlet for the Russian LEU while still guaranteeing customers a reliable, quality product at competitive prices.

USEC is pleased to perform this vital role in helping to carry out the national commitment to ensure that nuclear energy is used for peaceful and constructive purposes.

BACKGROUND

On February 18, 1993, representatives of the governments of the United States and the Russian Federation signed a government-to-government agreement to convert highly enriched uranium (HEU) recovered from Soviet-era nuclear weapons into low enriched uranium (LEU) for use as fuel in commercial nuclear power plants. To implement this "megatons to megawatts" agreement, on January 14, 1994 the United States Enrichment Corporation (USEC), serving as the U.S. executive agent, contracted with Techsnabexport, the Russian executive agent, to purchase reactor fuel grade uranium derived from 500 metric tons of HEU over a 20 year period.

By way of background, USEC is a wholly owned government corporation created by Congress under the Energy Policy Act of 1992 to restructure the U.S. Department of Energy's uranium enrichment program. The Corporation produces and markets uranium enrichment services to more than 60 electric utilities that own and operate commercial nuclear power plants in the United States and 11 foreign countries. The Corporation is headquartered in Bethesda, Maryland and operates plants in Paducah, Kentucky and Portsmouth, Ohio.

USEC will market the LEU derived from weapons-grade HEU to commercial electric utilities to fuel nuclear reactors throughout the world. USEC is the world's leading supplier of enrichment services, and is uniquely capable of marketing LEU derived from HEU without disrupting the international market while guaranteeing customers a reliable, quality product.

This historic "swords into plowshares" agreement goes beyond nonproliferation. Through it, the United States is expediting the dismantlement of nuclear warheads in the Russian Federation while ensuring that highly enriched uranium is used exclusively for peaceful purposes. Instruments of war are being converted into a valuable commodity that can help to pay for the cost of disarmament.

Under the contract, highly enriched uranium extracted from the equivalent of 22,000 Russian and Ukrainian nuclear warheads is being converted to low enriched uranium suitable for commercial power reactor fuel. Blending down of the HEU takes place in Russia, after which title to the low enriched uranium product passes to USEC. USEC is responsible for transporting the material to the United States. Once HEU is transformed into LEU, it cannot be made weapons-usable without going through the difficult enrichment process again.

The 500 metric tons of HEU that USEC will purchase over the next 20 years will be converted to 15,260 metric tons of LEU according to the following schedule:

- 6 metric tons of HEU in 1995
- 12 metric tons of HEU in 1996
- 10 metric tons of HEU per year, 1997 through 1999
- 30 metric tons of HEU per year for the remainder of the 20 years

The purchase is valued at $11.9 billion. The total transaction is equivalent to three years of world demand for enriched uranium. The quantity purchased in the first year alone will result in enough fuel to service 15 reactors and provide electricity to 10 million households.

Russia has committed to use a portion of the proceeds from this sale for conversion of defense enterprises, upgrading the safety of its nuclear power plants and environmental cleanup. Russia will compensate Ukraine for uranium from its strategic nuclear warheads by providing fuel for civilian power plants.

To ensure that the processed LEU is derived from HEU obtained from dismantled weapons, the two countries have signed a transparency agreement. The agreement states that Russia shall ensure that the HEU blended down to LEU is extracted from nuclear weapons. It also states that the Unites States shall ensure that the LEU received from Russia is used solely to fuel commercial nuclear reactors. The obligations of each party are subject to inspection and verification by the other party.

THE CONTRACT IS WORKING

Shipments were expected to begin in 1994. However, the contract specifies that the blended material must meet nuclear industry standards for purity. Initially, Russia had difficulty meeting those specifications. This delayed the initiation of shipments. USEC provided assistance to Russia to help solve the purity problem and in October 1994, Russia completed test runs confirming their ability to meet purity specifications. In December 1994, USEC placed its first order for six metric tons of HEU, the maximum amount that Russia indicated it would be able to supply, for delivery in calendar year 1995, USEC provided all shipment hardware, including product cylinders, sample containers and shipping containers.

USEC also took steps to ensure that proper precautions were taken to provide safe transport of the material from Russia to the United States. An environmental assessment was completed that determined that the transport of the blended-down HEU would have no significant impact on the environment. Low enriched uranium is a commercial product that has been shipped around the world by air, water and land transport for over 30 years. There has never been an accident involving the release of processed uranium during transport, and there has not been a transportation accident in the last 30 years that has resulted in a fatality or injury due to either the chemical or radioactive nature of processed uranium.

On June 23, 1995, the first shipment of 24 metric tons of low enriched uranium arrived at the Portsmouth Gaseous Diffusion Plant in Ohio. A total of nine shipments of LEU derived from 6 metric tons of HEU with a value of $145 million are scheduled during 1995. To date, eight shipments have been received at USEC's Portsmouth plant. The last shipment is expected to leave St. Petersburg next week. Agreement has been reached with Russia on both price and quantities for delivery in 1996. At Russia's request, USEC has ordered 341 metric tons of LEU derived from 12 metric tons of HEU next year.

USEC has paid Russia $160 million in advance payments to date; $60 million in four equal installments of $15 million from April through June 1994 and $100 million in July 1995. The $60 million was to be credited against deliveries in 1995 and 1996 and the $100 million against deliveries in 1996 and 1997. USEC has also credited half of the separative work unit (SWU) value in each delivery against the advance payment with cash payments for the remaining half. Accordingly, USEC has paid Russia $39 million for the first eight deliveries made in 1995, in addition to advance payments of $160 million.

The contract provides for separate payment for the enriched and natural uranium components, with payment for the enriched portion (SWUs) upon receipt and payment for the natural uranium due upon use or sale by USEC to third parties. Russia has indicated that it wants to change the terms of the contract to accelerate payment for the natural uranium component, which has an estimated value of $4 billion or approximately one-third of the value of the contract, based on a price of $28.50 per kg of natural uranium. Currently, the use of uranium for overfeeding is not economically feasible. U.S. antidumping restrictions inhibit the sale of natural uranium derived from Russian LEU to U.S. customers, Europe has also set import limits and Far East markets are not interested in purchasing Russian uranium.

Congress has proposed a solution to the problem of reselling Russian uranium that would open the U.S. market on a gradually increasing schedule starting in 1998. This is supplemented with a U.S. government commitment to buy all Russian uranium delivered under the HEU contract in 1995 and 1996 and hold it off the U.S. market until 2003. Several private sector firms have expressed support for this solution along with a strong interest in purchasing the Russian uranium delivered under the HEU contract.

Although the uranium issue remains unresolved, in a protocol dated June 30, 1995, both countries recommitted themselves to the sustained execution of the contract, agreeing to take steps necessary to advance its full and timely implementation.

Chapter V

International Management of Nuclear Power Fuel Systems

THERMAL ISSUES WITH THE U.S. HIGH-LEVEL WASTE REPOSITORY AND THE POTENTIAL BENEFITS OF WASTE TRANSMUTATION

Gordon E. Michaels

Oak Ridge National Laboratory
P.O. Box 2009
Oak Ridge, Tennessee 37831

INTRODUCTION

Recent total system performance assessments for the proposed Yucca Mountain site[2,3] (which assumes that spent fuel is directly emplaced in the repository without reprocessing and transmutation), have substantially altered the technical picture upon which much of the previous repository assessments were based[11]. The intention of this paper is to provide a qualitative update of the thermal issues associated with the proposed U.S. high-level waste repository and to discuss the potential impact of these issues. Significant questions about the ability to license the Yucca Mountain site are envisioned due to the difficulties associated with predicting the perturbations to the site that arise from the decay heat. It is suggested that waste transmutation, i.e., fuel reprocessing and use of Pu and other transuranic elements as fuel, may provide very significant benefits to the repository by removing the long-term heat source posed by actinides.

Actinides as a Source of Repository Heat

It is a standard assumption in repository assessments that spent fuel is emplaced after cooling times of 10 years after irradiation. At 10 years, approximately 77% of the thermal power in spent fuel originates from the decay of ^{137}Cs and ^{90}Sr and their daughters, with almost all of the remaining heat from actinide decay. In the short term, such as cooling times less than 50 years, the decay of ^{90}Sr and ^{137}Cs continues to produce the majority of the overall heat from spent fuel. However, over periods longer than 100 years, it is the alpha decay of various actinides, principally Pu and Am, that dominates the total heat output of spent fuel. Figure 1 shows the calculated total heat energy (that is, the thermal power integrated over time) for LWR spent fuel with the actinides present and with the actinides removed. The actinides account for about 80% of the heat that is generated over the first 1000 years. For time frames beyond 1000 years, actinides contribute about 99% of the additional heat generated in a repository.

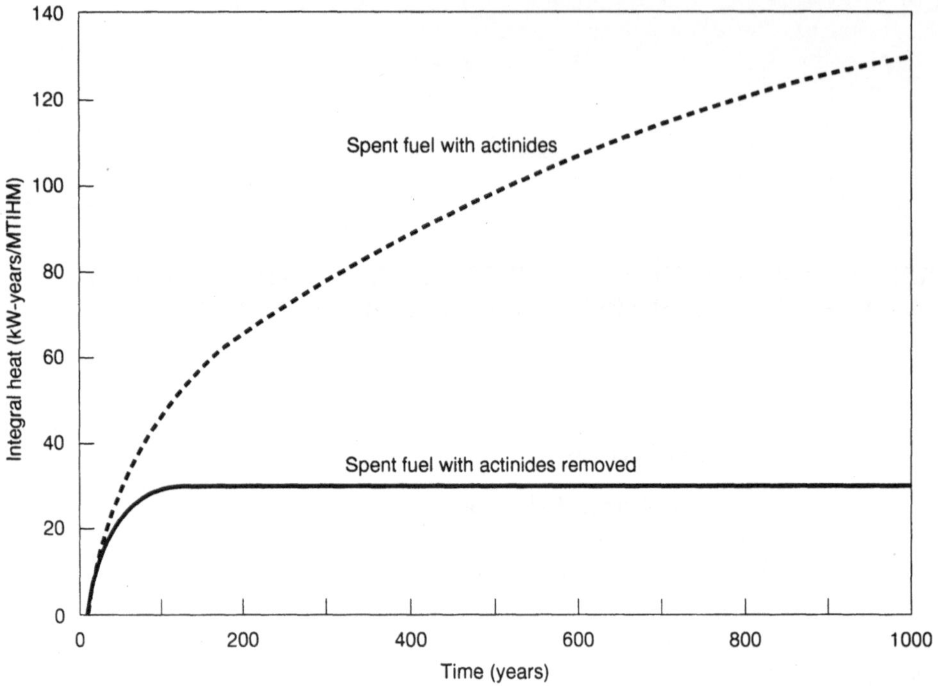

Figure 1. Total integral heat to be accommodated by a geologic repository. Heat values were calculated with ORIGEN2 model and are based on PWR fuel with an assumed burn-up of 30 GWd/MTIHM and 10 years cooling.

These longer time frames are important, because heat is expected to move relatively slowly through the geologic media. Bulk rock temperatures in and around the repository at any given time will depend upon the heat produced over long periods of time. As an example, rock temperatures at the so-called far-field location in the geologic media (about 50 meters from the emplaced waste) is not expected to reach its peak value until 700 years after emplacement of spent fuel[1]. Thus, in the context of overall bulk temperatures of large volumes of the repository site, it is the decay heat of the actinides in spent fuel that is the dominate heat source, not the decay of fission products such as Cs or Sr.

The Effect of Decay Heat upon Yucca Mountain

The proposed repository at Yucca Mountain would be at an elevation of about 200 meters above the water table but at about 300 meters below the surface of the mountain. Standardly[4,5] water is envisioned as contacting the surface of the mountain through rainfall or snowmelt and moving slowly downward through the repository region until it percolates into the aquifer, or water table, below. The standard scenario for releases from the repository involves water contacting the waste packages and accelerating their failure rate, followed by the alteration of the spent fuel by water, the subsequent dissolution of radionuclides within the groundwater, and the transport by aqueous pathways to the water table, where it is then assumed to have been released to an environment accessible to humans. The rock of Yucca Mountain above the water table has a moisture content that does not fill 100% of the rock pore volume; thus, it is said to be unsaturated. The relatively low level of moisture in the rock (70-90% saturated) and the slow speed at which water is able to move through the rock to the accessible environment are aspects of the Yucca Mountain site that were originally thought to contribute to the minimization of radionuclide releases for any given emplacement period.

However, the existence of the decay heat from spent fuel is now expected to substantially perturb the unsaturated zone of Yucca Mountain. It is now reported[2,3,6-9] that the hydrology and geochemistry of the site will be dominated by the heat of the spent fuel. Given the thermal-loading densities of spent fuel now considered for Yucca Mountain, a significant volume of the mountain will be raised in temperature above the boiling point of water. Thus, buoyant gas phase convection of water vapor will become an important feature of groundwater movement and of heat transport. Heat-driven buoyant vapor will transport water to a cool region above the emplaced waste where the water will condense and then be driven by gravity downward toward the repository. The resulting mountain-scale convection cells between the hot, above-boiling repository and the cooler, subboiling regions will create thermo-hydrologic phenomena not currently found in the ambient-temperature site. The mountain hydrology will change significantly over thousands of years as the mountain rock first heats up after waste emplacement and then cools because of the long-term reduction of the decay heat source term.

A schematic of the potential heat-driven effects upon the Yucca Mountain hydrology at the mountain-scale is shown in Figure 2. This figure is modified from Figure 10-1 of Reference 2 and is intended to show a physical picture of the thermally-driven hydrology as it has been developed by the U.S. Yucca Mountain project. The schematic presents a snapshot in time of the mountain-scale phenomena. Some of the relevant aspects are:

- A saturated zone above the repository will arise from condensation of thermally-driven buoyant water vapor. This is often called the condensation "perch" of the repository. The size and characteristics of the perch will vary in time and spatially, according to the spatial inhomogeneity of rock characteristics and of the thermal conditions. The saturated condensation zone may be tens of meters in depth and exist over hundreds of acres within the rock.
- Depending upon assumptions, the so-called dryout zone will only include the center portion of the emplaced waste, thus exposing the edges of the repository to near-boiling liquid water. Accelerated degradation of the waste packages from subboiling liquid condensation in the end regions of the repository is a concern.
- The so-called dryout zone will probably not be absolutely dry. First, the elevation of temperatures above boiling simply converts liquid phase water into the vapor phase; but, the resulting steam will leave the above-boiling region rather slowly[6,10] under most conditions and assumptions. Thus, the initial effect of heat is to create a hot (above-boiling), humid environment in which liquid film formation on the waste packages could lead to their accelerated failure. Additionally, aqueous phase water that is shed from the condensation perch will travel via fractures underneath the dryout zone and may be transported by heat pipe refluxing back into the dryout region, thus providing a mechanism for rewetting of the packages.
- A variety of possible phenomena could cause transport of liquid phase water from the condensation perch back into the waste package region. Instabilities in the surface of the condensation cap, arising from geologic or heat source heterogeneities, may cause

fracture flow of liquid phase water into the repository or enhanced imbibition of water into the local rock matrix (see Figure 2). Heat pipe effects may also affect transport of water around the condensation cap.

- The elevation of the rock to temperatures of 130°C or higher may cause alteration of the structure of fractures within the rock. This is important, because flow through rock fractures is currently believed to be the dominant transport mechanism for groundwater.
- Convection cells may arise within the water table. The geochemistry of the aquifer may be affected.
- The pH and chemistry of the groundwater may be greatly altered by the increase in temperatures due to the strong dependence of dissolution rates and solubility upon temperature and due to the introduction of flowing steam through the rock matrix.

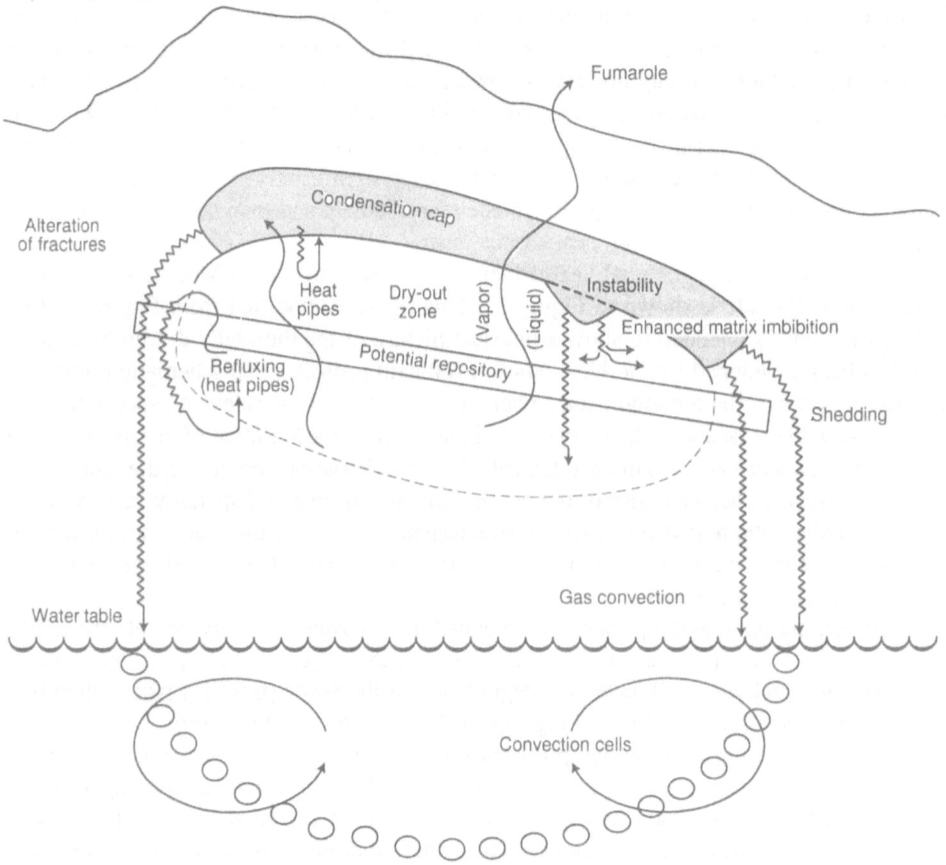

Figure 2. Schematic of thermally-driven processes that are expected to affect the hydrology of Yucca Mountain. Heat of decay of the actinides is the predominant source term that drives these processes. Some of the phenomena depicted are not yet included in repository performance calculations. Note that the so-called "dry-out zone" will actually retain non-zero levels of liquid saturation/humidity for long periods of time. Based on Figure 10-1 from Reference 2.

In addition to these mountain-scale phenomena, there will exist potentially important heat-driven phenomena at the spatial scale of the waste packages. During heatup and cooldown of the repository region, liquid phase convection cells can arise between neighboring waste packages, greatly altering the local groundwater hydrology. Additionally, Buscheck and Nitao[6] have developed a conceptual model of heat-driven vapor and condensate flow between two neighboring packages that are in regions of sharply varying bulk permeability; i.e., one package is in relatively unfractured impermeable rock and the second package is in rock that is significantly fractured. In the Buscheck-Nitao model, vapor-phase pressure differentials drive vapor flow into the fractured rock region where it condenses and drains, possibly causing persistent two-phase refluxing conditions that may "drench" the waste package. Such phenomena may generate a need to characterize the heterogeneity of fractures in the rock over the approximately thousand acres of the repository for the 10,000 to 30,000 waste packages.

Other temperature dependent phenomena important to repository performance will also be driven by repository heat. These include: solubility of the radionuclides, rates of corrosion pitting and microbial attack of the waste package, oxidation of the spent fuel, and transport time of $^{14}CO_2$ through the repository. An example of the latter consideration is shown in Table 1.

Given the 5450 year decay half life of ^{14}C, the difference in transport times of $^{14}CO_2$ from the packages to the accessible environment may be significant because of the potential for decay of the radionuclide during transport.

Table 1. Retarded Travel Times of ^{14}C from the Repository to the Atmosphere

Peak Bulk Temperature of Repository at Release Time	Average ^{14}C Travel Time to Atmosphere*	
	tuff permeability = $10^{-11} m^2$	tuff permeability = $10^{-12} m^2$
~50°C	1250 years	11,000 years
~130°C	250 years	4,500 years

* Values of travel time are derived from Figures 12-5 and 12-8 of Reference 2. Temperature values were obtained from Figure 12-6 of the same reference.

Implications of Actinide Heat for Repository Licensing

Licensing of a hot repository will require extensive analysis and characterization of a massively perturbed geologic system. Essentially, site characterization studies can only characterize the properties, hydrology, and chemistry of the unperturbed system. The licensing effort will need to rely upon computer models to extrapolate the existing mountain system to the massive-perturbed system that must be relied upon for isolation of the wastes from the accessible environment. Small-scale experiments may be able to provide helpful information, but the spatial scale of many of the discussed phenomena would appear to make their experimental characterization impractical. Thus, computer modeling will be required as a licensing basis.

For repositories that are above-boiling or near-boiling in temperature, the fact that heat transport will likely be dominated by multiphase convective processes rather than by conduction[6] will require enormously greater complexity in calculations of temperature.

Similarly, the transport of water and radionuclides in these systems will be driven by nonisothermal, multiphase flow with critically important[7,8] nonequilibrium flow processes between the fractures and the rock matrix. The modeling of these phenomena will require theoretical advances in flow modeling[6].

Another aspect of the effects of actinide heat upon the repository is the potential importance of heterogeneity to the performance calculations. The requirements for characterizing the heterogeneity of the mountain are unclear, but may well be very costly, and might be impractical. The issue of heterogeneity is further complicated by the possibility of heat altering the properties of the media; i.e., by alteration of fractures.

The temporal nature of the thermally-driven perturbation to the mountain system also introduces unfavorable complexity into the licensing process. Calculations by Buscheck and Nitao[9] indicate that repository heat-driven changes in the hydrology can persist for more than 100,000 years. The licensing requirements are expected to call for characterization of the performance of the mountain over a time period of 10,000 years or longer.

The combination of these factors: (1) theoretically complex heat transport and flow phenomena; (2) the importance of heterogeneity; (3) the diverse spatial scales of the processes, and (4) the long time periods over which the phenomena must be described, may make the licensing of Yucca Mountain very complex. Indeed, the current licensing strategy of the repository may be characterized as relying upon unprecedented success at modeling of very complex phenomena for very long periods of time which are sensitive to small-scale heterogeneities.

In this context, it may be that destruction of the actinides by transmutation with either accelerator-driven systems or in reactors, in combination with longer surface cooling times for Cs and Sr, would permit the emplacement of waste in an ambient-temperature repository that might be licensed in a more rapid and cost-effective manner. This "cold" repository concept would need to be studied further to determine its benefits and to define the requirements for actinide burning. It seems likely that the actinides would not need to be completely eliminated from the repository, but that reduction in actinide quantity by more than a factor of 10, but less than a factor of 100, might be required. Thus, transmutation systems would not need to address every waste stream that is to be emplaced in the repository in order to achieve the benefits of a cold repository — but only the majority of the actinide-bearing streams. This simplification is a potentially important aspect of a transmutation-enabled "cold" repository concept.

The projected benefits of a "cold" repository may be fully achievable only if both far-field and near-field temperatures are maintained at subboiling levels. Thus, transmutation-enabled cold repository concepts may require surface cooling times for Cs and Sr that are longer than 10 years in order to meet near-field temperature limits. This may be a potential problem for the cold repository concept and needs further study. Surface storage time requirements up to 50 years may be acceptable, and indeed may match well the lag storage times that will be required anyway as transmutation systems are developed, constructed, and brought into operation. However, Cs and Sr storage times greatly in excess of 50 years may present institutional issues for a transmutation-enabled cold repository concept. Clearly, better quantification of cooling time requirements for Cs and Sr are needed.

Methods other than actinide transmutation may exist for avoiding the above-mentioned thermal complexity issues. These methods may include:

finding a repository waste emplacement strategy and configuration for which the repository performance is insensitive to the theoretically difficult aspects of heat transport and flow that are expected for the hot repository.

decreasing the areal loading density of waste to maintain repository temperatures below the threshold for buoyant gas-phase flow processes. This repository strategy must result in temperatures that are significantly below "subboiling," e.g., peak temperatures below 70°C. This option would have unfavorable impacts on repository capacity.

The discussion of these other options is beyond the scope of this paper.

CONCLUSIONS

Many potentially important features, processes, and events have yet to be included in repository performance models, and many model simplifications have not yet been evaluated to determine whether they underestimate radionuclide releases. For this reason, it is premature to dismiss the radionuclide releases from the Yucca Mountain site as being acceptably low. Quantitative evaluations of the omitted phenomena are needed to determine whether the proposed Yucca Mountain repository will pose an acceptable health risk to the public.

Licensing of a repository at the Yucca Mountain site may be very expensive and may be uncertain of success due to the substantial perturbations to the mountain hydrology and geology caused by actinide decay heat. The heat of the actinides may have the following implications:

heat transfer calculations will not be focused upon theoretically-simple conduction processes, but instead will address multiphase, convective processes that are sensitive to spatial variability in the heat source term and in the geologic medium properties.

groundwater transport calculations will not be based solely on currently observable figures-of-merit such as groundwater percolation rates. Instead, thermally-driven processes will dominate the hydrology, and the licensing process will need to extrapolate multiphase flow processes with critically important disequilibrium processes between fractures and the rock matrix. These calculations will likely be sensitive to spatial variations in geologic media properties over hundreds of acres and may depend upon ability to forecast thermally-induced alterations of fractures in the mountain.

these theoretically-difficult processes will vary in time, and thus modeling of the massively perturbed mountain system must be performed for periods of 10,000 years or greater.

Transmutation of actinides, in combination with longer cooling times for ^{137}Cs and ^{90}Sr, may create an opportunity to design a relatively high capacity repository with peak temperatures below 70°C. Such a cold repository concept requires significant additional evaluation to determine whether it offers significant benefits. However, transmutation-enabled cold repositories have an intuitive appeal, because their licensing may be more directly based upon currently observable characteristics of the site and because their licensing may not depend upon unprecedented success with modeling of very complex phenomena over very long periods of time which are sensitive to small-scale system heterogeneities.

REFERENCES

1. Croff, A. G., "A Concept for Increasing the Effective Capacity of a Unit Area of a Geologic Repository," *Radioactive Waste Management and the Nuclear Fuel Cycle* (June 1994).

2. Wilson, M., et al., *Total-System Performance Assessment for Yucca Mountain - SNL Second Iteration (TSPA-1993), SAND93-2675*, Sandia National Laboratories (April 1994).
3. Andrews, R. W., T. F. Dale, and J. A. McNeish, *Total System Performance Assessment - 1993: An Evaluation of the Potential Yucca Mountain Repository,* B00000000-01717-2200-00099-Rev.01, INTERA, Inc., Las Vegas, NV (March 1994).
4. Barnard, R. W., et al., "TSPA 1991: An Initial Total-System Performance Assessment for Yucca Mountain," *SAND91-2795*, Sandia National Laboratories, Albuquerque, NM (1992).
5. Pacific Northwest Laboratory, *Preliminary Total-System Analysis of a Potential High-Level Nuclear Waste Repository at Yucca Mountain, PNL-8444*, Richland, Washington, Westinghouse Hanford Company (January 1993).
6. Buscheck, T. A., and J. J. Nitao, "The Impact of Buoyant, Gas-Phase Flow and Heterogeneity on Thermo-Hydrological Behavior at Yucca Mountain," American Nuclear Society, *Proceedings of the Fifth Annual International High Level Radioactive Waste Management Conference*, Vol. 4, pp. 2450-2474, Las Vegas, NV (May 22-26, 1994).
7. Buscheck, T. A., J. J. Nitao, and D. A. Chesnut, "The Impact of Episodic Nonequilibrium Fracture-Matrix Flow on Geological Repository Performance," *Proceedings American Nuclear Society Topical Meeting on Nuclear Waste Packaging (Focus 91)*, Las Vegas, NV, September 30-October 2, 1991. Also, *UCRL-JC-106759*, Lawrence Livermore National Laboratory, Livermore, CA (1991).
8. Nitao, J. J., T. A. Buscheck, and D. A. Chesnut, "Implications of Episodic Nonequilibrium Fracture-Matrix Flow on Repository Performance," *Nuclear Technology*, Vol. 104, No. 3, pp. 385-402 (1993).
9. Buscheck, T. A., and J. J. Nitao, "The Impact of Thermal Loading on Repository Performance at Yucca Mountain," American Nuclear Society, *Proceedings Third International High-Level Radioactive Waste Management Conference*, Las Vegas, NV, April 12-16, 1992. Also, *UCRL-JC-109232*, Lawrence Livermore National Laboratory, Livermore, CA (1992).
10. Mishra, Srikanta, "Far-Field Thermohydrologic Calculations," Appendix A in *Total System Performance Assessment - 1993: An Evaluation of the Potential Yucca Mountain Repository,* B00000000-01717-2200-00099-Rev.01, INTERA, Inc., Las Vegas, NV (March 1994).
11. Michaels, G. E., "Potential Benefits of Waste Transmutation to the U.S. High-Level Waste Repository," in *Proceedings of the International Conference on Accelerator-Driven Transmutation Technologies and Applications*, Las Vegas, NV (July 25-29,1994).

MANAGEMENT OF RUSSIAN MILITARY PLUTONIUM

C. Pierre Zaleski

Centre de Géopolitique de l'Energie et des Matières Premières
University of Paris-Dauphine
Paris

SUMMARY

The objective of this paper is to propose and to discuss a solution which makes it possible to store, as quickly as possible, all weapons-grade plutonium no longer used in the Russian military program, in a way that makes diversion extremely difficult, and the re-use in weapons form by the Russian government difficult and visible.

Two main conditions apply to this solution. First, it should be achieved in a way acceptable to the Russian government, notably by preserving plutonium for possible future energy production uses; second, the economies of the total system shall be as good as possible and there should be no charge, or a very limited one for the storage of plutonium.

I will propose a solution already outlined in Ref.1 : to store plutonium in a specially designed fast reactor, or at least a specially designed fast reactor core. I will attempt to demonstrate that this solution compares favorably to other possible solutions, applying the criteria set out in the goal and the two conditions mentioned above.

In addition, this solution should have the following side advantages :
- utilizing available personnel and installations of the Russian nuclear military complex;
- providing possible basis for decommissioning of older and less safe Russian reactors;
- giving a quantitative experience of construction and operation of a series of sodium-cooled fast reactors;

It also, however, presents a major problem : the need for rather large capital investment, with the risk of not getting the appropriate return on investment due to the generally difficult political and economic situation of Russia.

REVIEW OF OTHER POSSIBLE SOLUTIONS

The solution of mixing plutonium with fission products and disposing of this mixture in irretrievable deep geologic formations would, in our view, not meet the first condition, that is, the approval of the Russian government. In addition, this solution also presents a not negligible net expense, and may be questioned from the nonproliferation point of view, as in the long term one cannot exclude diversion of plutonium even from supposedly irretrievable storage, and the separation from fission products and re-use of plutonium may not pose major difficulties.

Placing the plutonium in the retrievable storage under Russian national safeguards may also be not very economic because of the cost of safeguards and the degradation of plutonium quality during the storage period, but more importantly, it would not prevent any future Russian government from very easily acceding to the plutonium for military use.

A more promising solution may be to "burn" this plutonium in the form of mixed-oxide fuel in existing light water reactors, modifying its isotopic composition in addition to mixing it with fission products. Here, however, two options are to be considered. One is to export the plutonium to other countries, for example Germany or France, which already have programs for MOX use, the second is to burn it in Russia. It seems to me that the first option has the following drawbacks :
- the difficulty for the Russian government to accept this export, losing its sovereignty over it (Ref.2);
- a possible issue of public acceptance in host country, whose public may be reluctant to deal with a problem it considers outside its responsability.
- the issue of the value of this plutonium; presently, countries using MOX are not limited by availability of plutonium but rather by a shortfall in fuel manufacturing capacities and the availability of reactors licensed to burn MOX fuel. For example, Electricite de France has recently decided that plutonium will have a zero value on its books. In fact, the Russian plutonium will displace French civilian plutonium, which will have to be placed in storage, with accompanying costs and degradation in quality. Thereore, the Russian plutonium will have to bear this cost, and will have a negative value. This may not be acceptable to the Russian government.
- the substitution of the Russian plutonium for French civilian Pu would in fact also be contrary to present French policy of having a minimum quantity of separated Pu in storage.
- the proliferation benefit of replacing weapons-grade plutonium in storage with civilian Pu would have to be clarified, including the question of weapons-worthiness of plutonium depending on isotopic composition and the technical capabilities of potential proliferators, an issue on which there is no agreement (Ref.3).

In summary, it is true that burning Russian plutonium in foreign reactors, for example French reactors, may appear an attractive solution; indeed, EDF's 28 reactors potentially able to burn the plutonium (with 1/3 MOX core), 16 of which are already licensed to do so, could burn the 100 tonnes of available Russian military plutonium in less than eight years. However, the above-mentionned six drawbacks mean that it is not a practical solution.

Therefore, one should rather consider burning or storing Russian plutonium in Russia itself. Burning it in existing Russian LWRs is theoretically possible. However, only some of these reactors of present design may accept plutonium, and no one are licensed for this. In addition, it would take many decades to burn the 100 tonnes in existing Russian LWRs (Ref. 4).

Another possible solution which seems relatively attractive is to burn the plutonium in newly built reactors. Fast breeders, which are clearly better for this purpose than LWRs (Ref. 4 and 4bis), have been proposed as a possible solution. Indeed, by completing two BN-800-type reactors in Russia, construction of which was begun some years ago but frozen for lack of funding, in a once-through cycle one could transform into spent fuel all 100 tonnes of ex-military plutonium in 30 years of operation (Ref. 4 and 4bis).

CONCEPT PROPOSED : FNPSR OR CAPTURE

Finally, the concept that I am proposing, which I have called Fast Neutron Plutonium Storage Reactors (FNPSR) and has also been dubbed CAPTURE (Ref. 5), has as an objective to store a maximum quantity of plutonium in the reactor core itself for a relatively long fuel cycle duration.

To achieve this, I suggest to seek a reactor core for which the typical design objectives will be :
- large fuel rod diameter - about twice that used in present breeder designs;
- low specific power, in range of 1 MWe per 10-15 kg of Pu;
- high internal breeding ratio, in the range of 0.9-1.0; and
- long fuel residence time, in the range of 10-20 years.

These design goals are consistent with the idea that plutonium has pratically no value for the coming decades (there is an excess of Pu) and that in case of ex-military Pu, one may even envisage a fee being charged for its storage, giving it in practical terms a negative value.

If one achieves this 10-15 kg Pu per MWe, between 10 and 7 GWe of CAPTURE type reactors would be needed to store all 100 tonnes of Russian plutonium as soon as the reactors are built. Another solution may also be considered, i.e., to built only half this capacity and irradiate the 100 tonnes of plutonium in two batches. However, this brings us closer to the previous solution with BN-800 reactors, and the differences become less clear-cut...

Assuming construction of CAPTURE reactor series totalling some 8 GWe, after maximum burnup is achieved, in some 30-35 years (15 for development and construction, and 15-20 for in-core residence time), the following options would be available :
- exchange Russian denatured plutonium for U.S. weapons-grade Pu, and use the latter for the second fuel load of the CAPTURE power stations;
- reprocess the spent fuel and re-use the Pu in the same reactors;
- store the denatured Pu in safeguarded, retrievable storage for future use, and use civilian Pu for the second fuel loading.

After two or three fuel cycles in CAPTURE plants, when the ractors have reached the end of their useful life, the following options would be available :
- if nuclear energy is to be discontinued and no development of breeders considered at this time, one could dispose of Pu definitively in underground repositories, or if environmental and proliferation concerns dictate it, burn the Pu in fast burner reactors (cf. French CAPRA program, Ref.5).
- if nuclear energy is developed as a long-term solution, as is expected by some today (notably the French, Russian and Japanese governments), one could use the Pu in fast neutron breeder reactors.
- if the conditions are still not clear 55-60 years from now, one could store the plutonium in a new series of CAPTURE power stations. This would preserve all options, and the eventual decision could wait for a clearer context.

One of the important aspects of this kind of project is timing. The following schedule may be imagined for implementing the CAPTURE idea :
- 10 years for the development and validation of the concept, notably fuel irradiation experiments ; however, start of construction on the first reactors could be envisaged before all experimental results are in hand, for example, seven years after the start of the project.
- construction time of 5-6 years, with construction starts every year on three reactor units.

With these rather optimistic assumptions, the total project would take some 15 years, by which time all military plutonium will be stored safely in CAPTURE reactors. Indeed, there is an implicit assumption in this optimistic schedule that the series of CAPTURE reactors, which will be rather simplified version of BN-600/800 (with lower specific power, simpler fuel handling equipment) will not require construction of a prototype.

Given that dismantlement of plutonium warheads will likely follow dismantlement of ^{235}U warheads, and that plutonium would be needed for fabrication of first fuel loads for CAPTURE about eight years after the beginning of the project, the above schedule is rather consistent with safe disposal of military plutonium.

TECHNICAL ASPECTS

With a plutonium value that is zero or negative - indeed, the owner of plutonium may be obliged to pay a storage fee, covering, for example, storage and safeguards costs - fast neutron breeder cores must be re-optimized.

Therefore, the natural idea is to use as much plutonium as possible in each CAPTURE core, so as to store more Pu per MWe, and also to increase the diameter of fuel rods, thus decreasing the relative share of fuel fabrication in the total cost of a kilowatt-hour. It is also natural to try to increase fuel residence time in order to simplify fuel handling equipment without penalizing availability. It is therefore desirable to seek higher internal breeding ratios in order to allow longer residence time without large reactivity swings.

Meeting these objectives will lead to a higher Doppler coefficient, which plays a positive role in controlling power excursions, but also to a more positive sodium void coefficient, which has an adverse effect on safety.

The core will, therefore, have to be optimized to ensure overall safety. This can, for example, be achieved via a relatively flat core design, with appropriate upper plenum design ensuring that any sodium voiding of a core section will inevitably lead to sodium voiding of the corresponding section of the upper plenum, where the reactivity effect can be designed to be negative.

For reasons associated with core design safety, it seems reasonable to limit the size of the power plant. A logical size would be between 600 MWe and 800 MWe, as Russian technology for this size of breeder reactor plant is well-developed.

The BN-600 FBR has been operating every successfully for over 10 years, with availability that is not only among the best in the Russian nuclear program, but also among the best in the world, all type of nuclear plants considered. The plant's average availability over 10 years was 97,5%, with a capacity factor of 71%.
In addition, Russia has developed detailed projects for the BN-800 fast breeder reactor, directly inspired from BN-600 technology (Ref. 4). Therefore, with some cooperation from western Europe (Phenix, Superphenix, EFR projects) and Japan (Monju), Russian scientists and technicians should be able to design and build a safe 800-MWe fast neutron power station without too much development.

The above-mentioned core optimization and the development and fabrication of Pu-bearing fuel with large-diameter rods will require close collaboration with western countries, notably with France, which has more experience in Pu-bearing fuel than Russia.

Core and fuel aspect represent the most innovative aspect of the CAPTURE design, and will probably require the most R&D. In fact, the idea that the author expressed in November 1994 was taken up for study by A.A. Kamaev of the Institute of Physics and Power Engineering in Obninsk, Russia and presented in Cadarache, France, in June 1995 (Ref. 6). His conclusions are that if one limits the core dimension by the diameter of the BN-800 vessel, the FNPSR core could not achieve the design objectives proposed in this paper.

The limiting values that Kamaev found were : specific power of 7,4 kg of Pu/MWe, inner diameter of fuel element cladding, about 8,8 mm; fuel residence time of 4,1 effective years; and breeding ratio of about 0,84. However, in his conclusion Kamaev stated : "Achievement of FNSPR type reactor characteristic values is possible in the new reactor plant design developed on the basis of design solution used for the BN-1600 heat removal system. In this case, R&D work should be carried out to substantiate design solutions used for the first time in the practice of home nuclear reactor building industry".

This clearly shows that the design goals mentioned above are not irrealistic, but will probably need more R&D. In fact, in Kamaev's preliminary review, he did not consider the possibility of increasing the diameter of the reactor vessel beyond that of existing projects (BN-1600) and changing the design of the fuel handling system. His preliminary results show that it would be necessary to study a rather flat core, with even larger diameter, larger fuel volume fraction, larger fuel element diameter and probably larger section of subassembly. This pancake-type core, with low specific power (kilowatt per liter of core), evidently calls for much simplified design of reactor vessel, vessel closure, and fuel handling system, the last perhaps inspired by those used for light water reactors (Ref. 7).

The relatively large vessel diameter required for this core design should not present major construction difficulties or increased costs, as we are talking about a relatively short stainless steel vessel with thin walls (no need to contain high-pressure liquid). The closure, the internals and the fuel handling mechanism should be much simplified compared to the classic breeder design. There may, of course, be some difficulties with the design, as for example the problem of intervention on failed fuel assemblies. One would have to determine by study if it is possible to accept a long waiting time before such intervention, inherent in the type of fuel handling I am suggesting, considering the low probability of this kind of event.

Generally speaking, this simplified vessel - wide and short - and simplified closure with no rotating plug but rather a type of leaktight cell (Ref. 7), associated with BN-800-

type components - such as pumps, intermediate heat exchangers, steam generators - may hopefully be considered, with proper design and test efforts, as reasonably proven.

ECONOMIC ASPECTS

A rough economic evaluation of the potential of this project may be done in two ways.

One way of evaluation is to use the 1993 EFR (European Fast Reactor) study carried out by utilities and manufacturers from France, Germany and the U.K., which seems the most recent, serious and pertinent study of the subject, and to transpose it to Russian conditions.

This study (Ref. 8) shows that in a western European context, and assuming zero plutonium value, a serie of 1,500-MWe FBRs can be in the range of economic competitivity with LWRs, at least with some uncertainty margin.

The EFR design was not optimized for plutonium storage; therefore, some gains can perhaps be anticipated thanks to :

- simplified fuel handling equipment design (fuel handling every 15 to 20 years);
- much lower cost fuel cycle : the large fuel rod diameter makes it possible to produce more energy per rod (for example, four times as much), and the cost of fuel rod fabrication should not be very sensitive to diameter; and
- potential fees to be paid by plutonium owners (for storage function).

These gains should thus lead in the EFR context to a situation very competitive with LWRs. This result should be transposable to the Russain context, asuming well-managed construction of a series of identical plants (10-800-MWe units).

There may, however, be one penalizing point, namely, the 800-MWe size, which is suggested for core design and local pragmatic reasons (the existence of BN-600 and of a detailed project for BN-800). However, this size is only slightly smaller than the largest modern Russian LWR of 1, 000 MWe (VVER-1000). Therefore, the economic size-related penalty should probably be significantly smaller than the advantage due to the re-optimization of the EFR core (see above). In fact, there is a certain tendency in Russia towards smaller reactors (for example, VVERs of 500 MWe and 630-MWe are in current Russian plans); thus, any economic penalty stemming from relative reactor size could disappear or even be reversed.

The second way to evaluate the economic potential of the project is to use the Russian internal comparison. According to V. Kagramanian, head of laboratory of systems analysis of nuclear power at the IPPE (Obninsk), a highly favorable experience has been gained in Russia from fast reactors : the BN-600 has the highest load factor among the country's reactors, and the BN-800 design modified (BN-800M) meets the latest, more stringent safely requirements. The economics of this reactor is equivalent to that of the medium-size thermal reactors (VVERs) or fossil-fired power plant (Ref. 9).

This, combined with gains from core optimization, gives a very positive indication about the economics of the concept.

One can therefore expect the CAPTURE project to have a good potential for competitivity with LWR projects in Russia, that is, that it could produce electricity at a lower price.

FINANCIAL ASPECTS

The need for new electric power plants in Russia is quite evident. Even if the domestic demand is not growing, because of the developments of energy saving and the efficient energy use, the possibility of exports to neighboring countries and the need to replace older power plants, nuclear or not, justifies some new construction.

The very difficult economic situation of Russia, however, makes the issue of financing difficult. Therefore, if this project is to go forward, international financing seems necessary, at least for most of the investment required. It seems to me that the international community, and especially OECD countries, should be interested in facilitating safe storage of weapons-grade plutonium. Some of the OECD countries also should be interested in maintaining world expertise and increasing operational experience in fast neutron reactor power plants, and the entire international community in helping the Russian economy reconstruct.

A potential additional motivation for the potential lenders may appear if the Russian authorities accept to link the building of these new plants with decommisioning of older and less safe nuclear power plants.

This being said, the important question is how Russia can reimburse the money borrowed for these projects.

What will certainly reassure potential lenders is a contract expressing the reimbursement in a commodity exported normally by Russia which has well-established international value and is in demand in the lending countries, for example, natural gas.

The other advantage to link the reimbursement to gas exports is that Russia can consider that it saves gas when producing electricity with new plant, and reimburses only part of the saved gas.

Some back-of-the-envelope calculations help to indicate some trends.

A study by the French Ministry of Industry (Ref. 10) analyzes the costs of a kilowatt-hour produced, on the one hand, by a combined-cycle natural gas plant and, on the other hand, by a nuclear plant. The plants are assumed to be in operation in the year 2003, and the discount rate is 8% for zero inflation. The capital cost (amortisation and interest charges) of a 1,400-MWe LWR plant expressed per baseload KWH represents some 64% of the cost in France of natural gas necessary to produce one KWH in a combined-cycle gas plant operating in the baseload mode, asuming a low (conservative) hypothesis for gas prices at the beginning of the next century.

Assuming that the capital cost of an 800-MWe CAPTURE power plant will be 1,3 times that of a French 1,400-MWe LWR plant - the 30% extra cost being compensated by a cheaper fuel cycle - amortization and interest charges for such a plant would amount to 83% of the cost of the natural gas needed to generate the same number of KWH.

In fact, as some eminent Russian scientists have indicated (Ref. 11), in the present situation and for a given amount of hard currency, the Rusian nuclear industry may perform much more work than western industry. For specific examples related to the upgrading of old Russian reactors, they suggested that Russian industry would be 16 times more efficient than western industry.

To be conservative, it would not be extraordinary to consider a factor of two to characterize the relative efficiencies of the two industries (Russian and western). This may be due to the relatively low cost of labor in Russia expressed in hard currency, as well as the high contribution of labor costs in the total cost of nuclear plant construction.

Assuming further that at least 80% of the construction work on CAPTURE stations could and would be done by Russian industry, the cost, in hard currency and as a percentage of typical western costs for the same work, can be expressed as 80%/2 + 20% = 60%. In this context, it would suffice to devote 50% of the gas saved by operation of the nuclear plant to the payment of interest at 8% and to amortization of the capital cost.

On the other hand, the value of gas in Russia (where the nuclear plant would be built) is lower by some 35% than it is in France, due notably to transportation costs. Therefore, 80%, and not 50%, of gas saved during the 30-year economic reactor operation would have to be devoted to financial costs. This still appears a reasonably good deal for the Russians, as the fuel cycle cost for CAPTURE (which is very low) and its operations & managemet (O&M) costs should not exceed O&M plus amortization costs for an equivalent-size- gas-fired power plant.

As the Russians may have a problem to market all their available gas, it would probably be necessary to conclude a separate contract to purchase gas for reimbursement of the CAPTURE project, in addition to normal commercial gas supply contracts.

It is quite clear that financing is the largest obstacle to the entire CAPTURE project, however one can think that all governments interested in the main objective of CAPTURE - managing excess Russian military plutonium - as well as governments interested by the side benefits outlined at the beginning of this paper, would like to help with this issue.

Using as far as possible the expertise and available time of Russian scientists and engineers, as well as their installations, development expenditures for this project should not be excessively high. If the development phase is successful, construction of a series totalling some 8 GWe of CAPTURE reactors should not exceed about $ 10 billion. In addition, the project should be self-supporting, with no, or very modest, fee for military plutonium storage. The risks - political, economic and technical - may discourage normal financing. One can, however, imagine that the governments interested in the project may also subscribe some sort of guarantee for the risks involved.

CONCLUSION

The CAPTURE concept seems potentially sufficiently attractive to at least deserve further investigation. The next, relatively inexpensive step, could be the study of cores without the constraint imposed by vessel diameter. If this study leads to satisfactory results, a slightly more expensive, but still not too costly, step migh be contemplated, to study and design a simple, large-diameter, low-height, vessel with a very simplified closure, internals, and fuel handling equipment.

Thereafter, a decision should be taken if more costly steps involving mockups, tests, fuel irradiation, and other design and experimental efforts should be launched. It is so be hoped that at least the first step will be launched in the near future.

REFERENCES

1. Zaleski, C.P., November 1994, *Fast Neutron Plutonium Storage Reactors : An example of the utilization of the military nuclear complex for peaceful purposes.* UNESCO ROSTE, Venice.

2. Remarks of N. Yegorov, first deputy minister, Ministry of Atomic Energy - Russian Federation Proceedings of seminar on Back End of Nuclear Fuel Cycle, CGEMP, Université de Paris Dauphine, May 1993.

3. Remarks by C.P. Zaleski, annual conference of Japan Atomic Industrial Forum, Tokyo, Japan, April 1995.

4. Kagramanian V. et al., "Aida-MOX" Proceedings of Global 95, Versailles, France September 1995.

4bis. Mikhailov, V.N. et al., Ministry of Atomic Energy - Russian Federation, Moscow, V. Mugorov et al., Institute of Physics and Power Engineering, Obninsk, Russia; I.N. Avrorin et al., NNIITF, Moscow, "Plutonium in Nuclear Power of Russia". OECD-NEA Expert Group on the Management of Plutonium, Paris, March 1995.

5. Zaleski, C.P., 1995, "Fast neutron reactors : development in future decades" *International Journal of Global Energy Issues,* Oxford, U.K., Vol.8, Nos 1-3, pp. 133-142.

6. Kamaev, A.A. "Technological Aspects of FNPSR Development on the Basis of Russian LMFRs. "French-Russian seminar, Cadarache, June 1995.

7. *Interdepartemental Fast Neutron Reactor Studies.* Electricité de France, Direction des Études et Recherches, Clamart, France 1971-72.

8. Lefèvre, J., Hubert, G., and Aubert, M. "Le projet de réacteur rapide européen EFR: état actuel et perspectives." *Revue Générale Nucléaire* , No.6, p. 504- 516, Paris, 1994.

9. Ermakov, N.I., Ministry of Atomic Energy - Russian Federation; V.N. Murogov and V.M. Poplavski, Institute of Physics and Power Engineering, Obninsk, Russia. "Role of Fast Reactors in the Future of Power Engineering, Fuel Supply and the Environment." International Topical Meeting on Sodium-Cooled Fast Reactor Safety, Obninsk, September 1994.

10. Ministèrede l'Industrie, des Postes et des Télécommnications. *Les Coûts de Référence - Production d'Electricité d'Origine Thermique.* Paris, 1993.

11. Ponomarev-Stepnoy, N.N. and Adamov, E. in Proccedings of the MIEC-CGEMP seminar, *La Sécurité de l'Approvisionnement en Energie d el'Europe : Rôle de la Russie.* CGEMP, Université de Paris-Dauphine, Paris, 1994.

Chapter VI

Relevance of International Consensus Policies on Alternative National Energy Strategies as Related to Free Market Economic, Global Fuel Transportation Systems for Oil and Gas, Environmental Impact and Geopolitical Trends

TRANSPORTING OIL FROM THE CASPIAN SEA TO WESTERN MARKETS: A TURKISH PERSPECTIVE

Rafet Akgunay, Minister-Counselor/Deputy Chief of Mission

Turkish Embassy
1714 Massachusetts Avenue, NW
Washington, DC 20036

INTRODUCTION

It is a given fact that the technological developments of the 20th century introduced new sources of energy other than fossil fuels. Nuclear reactors today, for instance, generate approximately 17% of the total worldwide electricity output.

Despite these developments, the world continues to remain substantially dependent on fossil fuels for the foreseeable future, until research and development can advance alternative fuel technologies. Apparently, changes in the pattern of energy consumption will be evolutionary rather than revolutionary.

Until such time comes that the use of renewable energy dominates the market, fossil fuels, consequently, will continue to be almost the sole fuel mix, particularly in the transportation sector. Furthermore, even though new technologies will make energy consumption more efficient in this sector, such gains might very well be offset by increases in traffic of all types.

Under these circumstances, the security of oil resources and the transportation of oil to world markets will continue to be of utmost importance. In this framework, a central requirement for future energy security will be to assure the diversification of the oil supply, especially from sources outside the Persian Gulf. In so doing, the domination of any one region of the world over international oil markets or oil supplies can very well be prevented.

Yet another crucial component in the security of oil supplies is the proper transportation mechanism. A politically stable, economically efficient and viable mode of transportation is the sine qua non for ensuring secure petroleum supplies and moderating world oil prices.

CASPIAN SEA BASIN

The Caspian Sea basin constitutes one of the world's foremost and, not to mention richest, oil repositories. Although exploration is still in its earliest stages, 42 billion barrels of proven oil reserves certainly offer exciting possibilities to the nations in the region as well as to the oil consuming countries of the West. More specifically, the region's newly independent states of Azerbaijan, Kazakhstan and Turkmenistan, which have few major industries, stand to gain economic and political strength and stability from the development of these vast national resources. This, in turn, would help them on their respective roads to true independence. For the West,

Economics and Politics of Energy
Edited by Kursunoglu *et al.*, Plenum Press, New York, 1996

on the other hand, the Caspian oil could very well help diversify the world's oil supplies, stimulate price competition, reduce monopolistic control and increase the security of energy supplies. In short, there is much to be realized by all parties involved.

In fact, the nations in the region have already embarked on aggressive efforts to develop these resources in partnership with the world's leading energy companies. Azerbaijan, for example, signed a $7.5 billion contract with a consortium composed of 12 energy companies from seven countries--namely, AMOCO (US), PENNZOIL (US), UNOCAL (US), McDERMOTT (US), EXXON (US), BP (UK), RAMCO (UK), LUKOIL (Russian Federation), STATOIL (Norway), TPAO (Turkey), DELTA NIMR (Saudi Arabia) and SOCAR (Azerbaijan)--to develop three oil fields over the next twenty years, with an estimated production of over three billion barrels of crude oil. The Kazakh Government also entered into a $20 billion joint venture with CHEVRON (US) to develop the Tengiz field over the next 40 years, with a potential yield of 6 to 9 billion barrels of crude oil. The Tengiz field is one of the world's ten largest oil fields and the largest to come into production in the last twenty years. In addition to these initiatives, SOCAR, LUKOIL, PENNZOIL and AGIP (Italy) signed a contract for the exploration of the Karabagh fields in Azerbaijan. They join other companies in the region in various stages of exploration or production.

TRANSPORTATION OF CASPIAN SEA OIL TO WORLD MARKETS

Even with the ongoing development of the oil of the Caspian Sea region, there remains one major problem--there is no outlet to Western markets. Clearly, the challenge confronting each of these nations and the involved energy companies is how they will transport the oil they produce to Western markets. Two alternative routes have been proposed, and are currently under consideration, for the long-term transport of oil from Azerbaijan. One is the Northern Route: a pipeline across Chechnya, through Russia, ending at the Black Sea port of Novorossiysk. This would then require shipment by tanker across the Black Sea and through the Bosphorus Straits in order to reach the Mediterranean and subsequently the West. The other alternative is the Mediterranean Route: a pipeline across Turkey, ending at the Mediterranean port of Ceyhan, totally bypassing the Black Sea and the Bosphorus.

THE BOSPHORUS CHALLENGE

In assessing the viability and feasibility of the proposed routes, it is crucial to analyze several factors. One of the most significant challenges to consider is that posed in using the Bosphorus. The Bosphorus is one of the world's busiest waterways, handling some 45,000 major ships annually. This is three times the traffic in the Suez Canal, and does not even include the thousands of crossings by local boats and ferries plus fishing and pleasure boats. Almost a half million residents criss-cross the Bosphorus daily on local ferries.

The Bosphorus is one of the world's most difficult waterways to navigate. Nineteen miles long and less than a half mile, or 700 meters, at its narrowest point, the Bosphorus has abrupt and angular windings that require ships to change course at least twelve times, including four separate bends requiring more than a 45 degree turn. On at least two of these sharp turns, ships approaching in the opposite direction cannot be seen. There are powerful and rapid currents of 5 to 8 miles per hour, variable counter currents and submerged eddies, capable of dragging ships off both course and anchor. This has caused collisions and groundings of ships, which sometimes even crash into buildings on the shore. There are additional hazards since two major bridges span the Bosphorus.

The situation in the Bosphorus becomes even more perilous when we consider the fact that traffic there is basically unregulated. Under the 1936 *Treaty of Montreux*, the Bosphorus was opened to all merchant ships of all nations. Unlike other comparable waterways, reporting on cargo content is voluntary, even with regard to hazardous cargo including nuclear, flammable or toxic waste. The use of pilots is also voluntary, even though it has been demonstrated to significantly reduce the risk of accidents. In fact, only one percent of collisions occurred when pilots were used on both vessels. Consequently, under the terms of the *Montreux Convention,* there is no mechanism for taking special precautions because of hazardous materials, for staging or scheduling traffic, or for requiring that pilots be used by foreign ships which are unfamiliar with the route. No other major international shipping waterway of this magnitude operates in such a manner.

Shipping traffic in the Bosphorus has grown dramatically and is likely to continue to do so, both in terms of numbers and size of ships. Since 1960, the number of foreign ships has increased by over 150%, while the tonnage of these vessels has increased by over 400%. To handle the Caspian Sea oil, a significant increase in tanker traffic would naturally be required. For example, if 45 million additional tons of crude oil were to be transported annually, it would require 4,500 small tankers of 10,000 tons each way--a 20% increase in total Bosphorus traffic and a total of 9,000 passages--or 450 additional large tankers of 100,000 tons each way for a total of 900 passages. However, in the past three years only seven ships of 100,000 tons have gone through the Bosphorus and with such difficulty as to require closing the waterway to other traffic.

It is also imperative to keep in mind that the Bosphorus cuts through one of the world's major cities, Istanbul, where over 12 million inhabitants live in proximity to both shores. Any shipping calamity involving fire, explosion, toxic or nuclear material could endanger the health and lives of millions of innocent people. Moreover, scientific studies reveal that the marine environment is already in great jeopardy from the heavy traffic, including ship waste and pollution. Oil tankers pose a particular danger from leakage, improper flushing and possible collisions, compounded by the continued use of many old tankers.

As a result of all these factors, it is evident that the Bosphorus is already an extremely hazardous waterway. 167 large scale accidents occurred between 1983 and 1993, and the average annual rate of accidents increased significantly since 1988. In February and March of 1994 alone, there were five collisions, including one involving the Greek Cypriot tanker "Nassia" which resulted in over 30 deaths, an oil slick that burned for five days and closed the straits to traffic, and the dispersal of 20,000 tons of oil both in the water and along the shore which severely damaged the ecosystem. A 1979 collision between two Rumanian and Greek tankers resulted in the release of over 95,000 tons of oil, some of which burned for weeks. In 1991, a Lebanese vessel, the "Rubinion 18,"struck one of the bridges and sank with its cargo of 20,000 live sheep. The resulting decomposition, as well as oil leakage, had noxious effects on sea life and the environment.

In short, because of the precarious situation of the Bosphorus--from an environmental as well as safety perspective--additional traffic would seriously exacerbate conditions. If the Northern Route to the Black Sea is selected, any major collision or accident that forces the Bosphorus to close for any period of time would disrupt all transport to Western markets of Caspian Sea oil sent via this route.

THE MEDITERRANEAN ROUTE: THE BETTER OPTION

Logically, the proposed pipeline from Baku through Turkey to Ceyhan on the Mediterranean--what is known as the Mediterranean Route--is the better alternative. It has already been demonstrated to be technically feasible. It would best meet the commercial needs of the energy companies and the nations involved as well as the geostrategic interests of the West. Several factors support this conclusion.

First, Ceyhan already has a modern, state-of-the-art tanker loading and storage facility and needs no additional construction. It was built to accommodate the Iraqi oil pipeline, which Turkey shut down in 1990. Ceyhan can handle the largest tankers in service--300,000 tons--far larger than the size that can navigate the

Bosphorus. It has four times the capacity of Novorossiysk. Since Ceyhan is on the Mediterranean, the dangerous passage from the Black Sea, through the Bosphorus and the Dardanelles is eliminated. Ceyhan can also operate 365 days a year, compared to the more difficult meteorological conditions in the northern Black Sea at Novorossiysk which on average loses as much as two months of the year to adverse weather. The Turkish Government's proposal takes ultimate responsibility for financing and completing a Ceyhan pipeline.

Second, since Ceyhan delivers oil directly to the Mediterranean, eliminating the difficult and expensive passage through the Bosphorus, the lowest international shipping rates to key cities in Western Europe would be available at approximately half the estimated cost from Novorossiysk. The Turkish proposal in this regard also includes guaranteed competitive, non-discriminatory tariffs and access. The Mediterranean Route is far more competitive.

Third, as mentioned previously, by going directly across Turkey to the Mediterranean, the Ceyhan pipeline would avoid the Bosphorus and, thus, eliminate the added danger of collisions that would result from greatly increased ship traffic and that would further endanger the residents of Istanbul and the waterway's delicate eco-system.

Fourth, because Ceyhan is located in a NATO nation on the Mediterranean, it would provide maximum security for Western consuming nations. Any shipping accident in the Bosphorus, which is always possible, would therefore not effect the continued flow of oil to the West if the Mediterranean Route is selected.

In the end, the evidence clearly proves that the Mediterranean Route is the best commercially and the best environmentally for transporting Caspian Sea oil to Western markets. Turkey fully supports this alternative. While the Ceyhan pipeline would surely meet the long-term transport needs of the Caspian oil fields, the Azerbaijan International Operating Company--an international consortium of 12 energy companies known as AIOC--is not scheduled to select a route for the main oil until next year. Subsequently, it will then take several years to complete construction of the pipeline.

EARLY OIL

However, there is an immediate need for a transport route to carry the first oil--the "early oil"--produced in the Azeri off-shore oil fields. Given this need, the Turkish Government has submitted to the AIOC a detailed proposal to transport this early oil via the Western Route--that is, across Azerbaijan and Georgia to the Black Sea port of Supsa. Turkey is willing to refurbish an existing pipeline between Baku and Supsa and to construct the missing links.

To make this route economically feasible and competitive, the Turkish Government has provided guaranteed attractive tariff levels and guaranteed concessionary financing to the AIOC, if needed, in case the AIOC decides to build the pipeline itself. If the AIOC decides otherwise, Turkey asked that it be authorized to build and operate the pipeline under specified tariff terms and conditions. Turkey would create this project company with the participation of the involved governments in Azerbaijan and Georgia and has invited the Russian Government to participate as an equity partner as well. The alternative is shipment of the early oil to Novorossiysk, following the route of the main oil.

Turkey has offered to purchase all early oil, which is expected to be around 6 million tons per year, for its own domestic use at market rates. Since Turkey currently imports 20 million tons of oil annually, it can readily use this oil to the mutual benefit of all parties in terms of economics, efficiency and safety. Turkey has committed to spend extra funds to transport this oil to its own refineries while avoiding the Bosphorus.

The AIOC decided on October 9, 1995 to use both routes for early oil transport and declared that the final aim of the project will be to reach Ceyhan. Turkey believes that this option has the best commercial terms and is the best environmentally. Use of the Western Route would also provide the newly independent states of Azerbaijan and Georgia with the maximum opportunity to participate in this project, thereby strengthening and stabilizing their economies and political independence.

TURKEY, US INTERESTS AND RUSSIAN PARTICIPATION

The US, for its part, has been following this issue rather closely. Turkey and the US find themselves on the same track. Without a doubt, the United States--and for that matter, the West in general--have several key points of interest in this decision. First, tapping into the Caspian Sea oil reserves would encourage the diversification of oil suppliers for Western oil-consuming nations. This would naturally prevent any one nation from monopolizing the vast new Caspian Sea region oil reserves. Second, it serves the West's interests in ensuring the flow of oil along secure routes, with minimal risk of disruption, since such routes would come under the control of countries that are allied with NATO and the Western defense network. Third, since many companies play a dominant role in this process--especially with CHEVRON as the lead in the Kazakh oil field and five US companies as the largest national grouping in the AIOC, owning a combined 43% of the equity--it is important that they are offered fair commercial terms, given open access and non-discriminatory tariffs. After all, these corporations are committed to investing tens of billions of dollars in exploration and production. Fourth, it is also in the interests of the international community to assist the newly independent states around the Caspian Sea to strengthen their economic and political systems and to continue their transition from Soviet domination to stability and true self-supporting independence. Last, but not least, the pursuit of this initiative ought to encourage maximum protection of the environment and safety in all aspects of energy development and transport.

The United States should be--and has been--particularly sensitive to the position of Turkey, which shut down the Iraqi pipeline terminating at Ceyhan at the request of the United Nations after Iraq's invasion of Kuwait in 1990. Since then, the Iraqi pipeline has remained closed, the modern tanker terminal at Ceyhan has sat idle, and Turkey has suffered the loss of billions of dollars in fees. None of the Allies in the Gulf War continues to suffer as great a loss in revenue. The early oil route to the West and the Ceyhan pipeline present an opportunity, on a fully competitive commercial basis, for Turkey to again participate in the international energy transportation industry. Turkey is ready and willing to seize this opportunity, in cooperation with our neighbors in the region, energy companies as well as the US.

This includes cooperation with Russia. The importance of Russia in this region cannot be overlooked. Turkey very much favors the Russian Federation maintaining its unity and territorial integrity in a politically and economically stable environment. The Russian Federation can also become a positive influence and a factor of peace and stability beyond its own borders. In Turkish-Russian bilateral relations, Turkey considers Russia as today's and tomorrow's partner, pursuing regionally and mutually beneficial endeavors. In this instance, seeing Russia in a partnership with the West, reaching Ceyhan together in the proposed oil pipeline project, is not a fantasy. It is very possible and a reality worth realizing. Developing a common understanding and a mutually beneficial cooperation in a partnership of this nature will set the groundwork not only for the achievement of the oil pipeline project but also for securing peace and stability in the region and the world at large.

DISCLAIMER: The views expressed in this article belong solely to the author.

RENEWABLES: A KEY COMPONENT OF OUR GLOBAL ENERGY FUTURE

Dan Hartley

Vice President, Laboratory Development
Sandia National Laboratories
Albuquerque, New Mexico 87185

INTRODUCTION

Inclusion of renewable energy sources in national and international energy strategies is a key component of a viable global energy future. The global energy balance is going to shift radically in the near future brought about by significant increases in population in China and India, and increases in the energy intensity of developing countries. To better understand the consequences of such global shifts in energy requirements and to develop appropriate energy strategies to respond to these shifts, we need to look at the factors driving choices among supply options by geopolitical consumers and the impact these factors can have on the future energy mix.

There is no argument that both today and in the future, fossil fuels will account for the lion's share of energy production. The EIA puts world primary energy production numbers at: fossil fuels - 86.6%, renewables - 7.0%, nuclear - 6.4% (EIA 1993). Of the renewables portion, over 95% is hydropower. The EIA predicts that renewables will grow at a faster rate over the next twenty years (2.3% annual rate of growth) than will energy production overall (1.6%) (EIA 1995).

Geopolitical energy consumption patterns also will shift, with significant economic and political consequences. Predictions are that the world's population will increase by over 30% in the next 25 years (an increase of 2.7 billion); almost two-thirds of that growth will be in Asia and Latin America (*The Economist* 1994). China alone is predicted to grow between 100 to 200 million in that period, accounting for up to nearly 10% of that growth (Lu 1993). Energy production in China alone has shown phenomenal increases over the last 30 years, increasing 1,850%, making China the third-largest national producer in the world (Lu 1993).

Key to our argument is the proposition that a 60 quad increase in energy consumption in the industrialized countries has significantly different global consequences in a variety of dimensions than does the same-sized increase in the non-industrialized world. The integrated nature of the global geopolitical community as well as the magnitude of the ecological impact development activities may have suggests that energy production activities anywhere in the world will have consequences throughout the global system. The predicted growth in energy consumption in the developing world thus has particular implications for energy strategies in the industrialized nations - implications which, we suggest, warrant serious consideration and inclusion of renewables in any future energy strategy. An analysis of the factors driving choices among supply options for all countries and the global consequences of such national choices will support this statement.

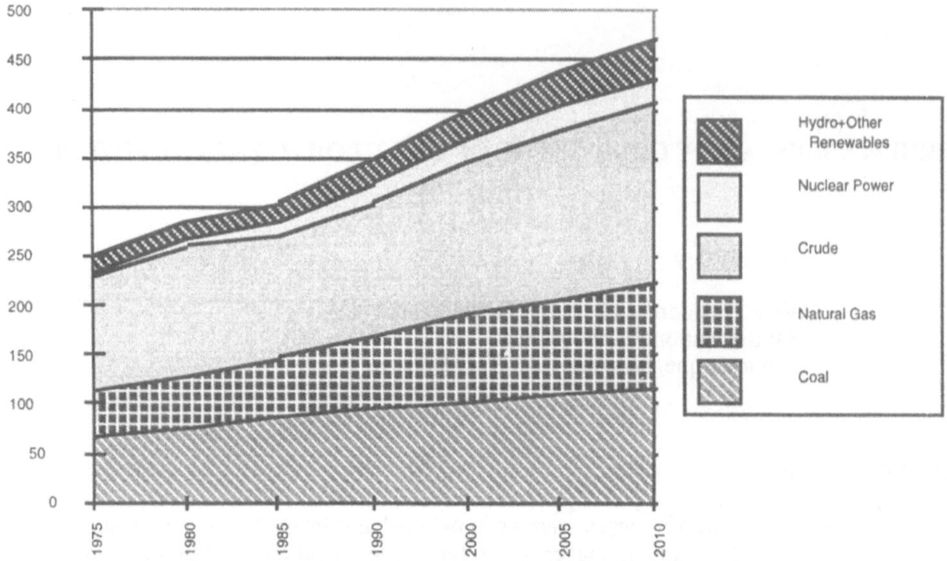

Figure 1: Global energy production in quads
source: EIA Annual Energy Review 1994 and EIA International Energy Outlook 1995

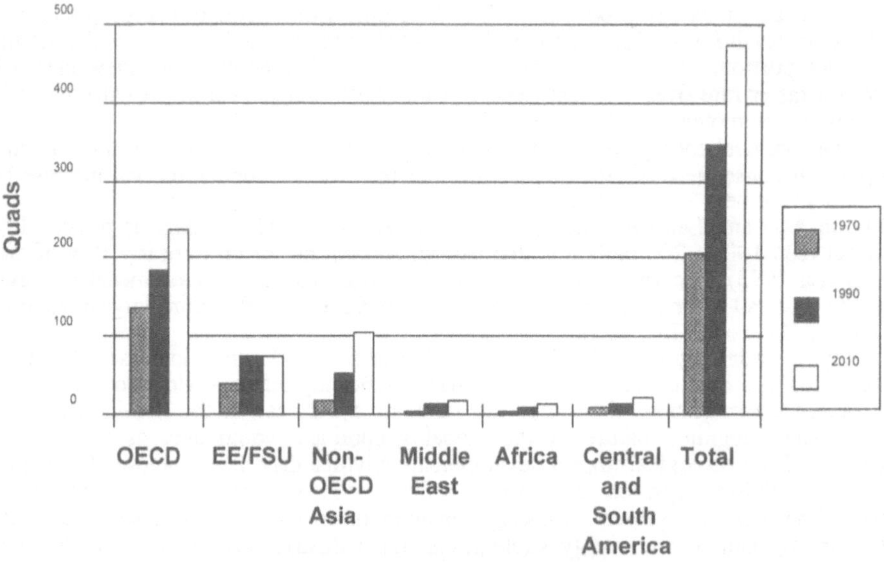

Figure 2: Global energy consumption
source: International Energy Outlook, 1995, EIA

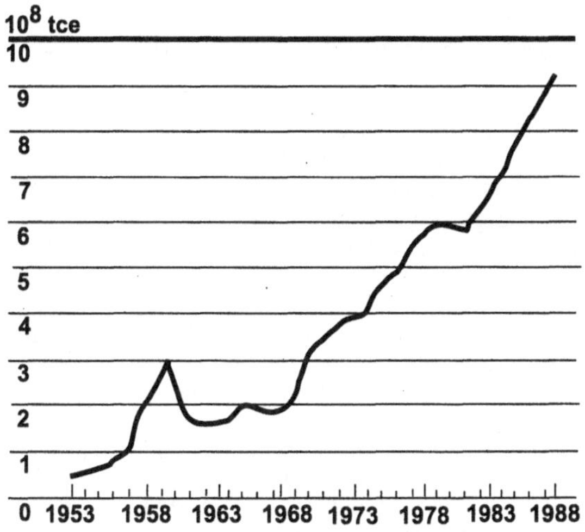

Figure 3: Energy production in China over time

source: Lu 1993

RELATIONSHIPS AMONG ENERGY SUPPLY OPTIONS AND FACTORS DRIVING CHOICES AMONG THEM

Energy Supply Options

Energy supply options generally fall into the following categories: fossil fuels, civil nuclear power, renewables, conservation, and new 'unconventional' sources under development. Fossil fuels can be further divided into oil, used primarily for transportation, coal, and natural gas. While prices for oil are generally seen to be going up (see, e.g. EIA, IEA, and DRI:McGraw Hill predictions in US Department of Energy, Energy Information Administration, 1995), natural gas is getting cheaper and coal is seen as a readily available and relatively inexpensive energy source. Civil nuclear power is under significant challenge in the US (no new plants are under construction and licenses for current plants will begin to expire around 2010 with controversy regarding extension and renewal), but is seen as a viable new source in developing countries such as China and Indonesia, and is well-established in some industrialized nations such as Japan and France. In the renewables set, we typically assume that this means solar. In fact, hydropower is a well-accepted and relatively inexpensive form of solar energy production and has the greatest commercial usage of the renewables. Hydropower is followed in usage by biomass, geothermal, and wind in that order. Solar power technologies (thermal, photovoltaic) are just beginning to penetrate the commercial world. Conservation has a varied history; as demand-side management and energy-efficient technologies it has gained credibility over the last decade in the US, overcoming some major institutional and infrastructure barriers (such as utilities' pricing schemes) to do so, while in China it has been called "The Fifth Energy Source" (Lu 1993). 'Unconventional' sources such as hydrogen and nuclear fusion are still under development; predictions as to the timing of their emergence from the laboratory to the commercial sector vary widely. In this paper we will focus on the first three supply options (fossil, nuclear, and renewable energy) as conservation is a means for obtaining energy services of a qualitatively different sort (focusing on changes in behavior and technology replacement) and the unconventional sources are too far from market to be commercially feasible.

Factors Driving Choices Among Energy Supply Options

We propose to address choices among these energy supply options in terms of five major factors that drive the decision process: economics; energy self-reliance or independence; the existence of various types of international agreements; the existence or absence of an established infrastructure; and the relationship of the choice to sustainability.

Each of these factors has several dimensions, and the application of and weight applied to each factor significantly impacts the world energy mix with implications for global quality of life.

Economics. The economics factor has two primary dimensions: the levelized cost of energy, or the cost per kilowatt-hour to the customer (kwh), and the ease of obtaining financing for energy projects. The cost per kwh is relatively self-explanatory. The ease of obtaining financing for energy projects is critical to developing countries, for they often have significant needs for grid development as well as capital construction, and cannot finance these out of internal cash flows. The picture is complicated as, in general, international lending practices preclude rural electrification (the biggest area of need in developing countries, including China and India) because the return on investment requirements placed by lenders demand an early and substantial cash flow from electricity generation to meet loan payments (cf Mellecker 1995), and political considerations favor those projects which serve the greatest number of people in the shortest amount of time, again favoring grid-connected applications. This type of financial structure favors grid-connected consumers, i.e. urban or developed areas, which will generate quick cash flows, and technologies with low up-front (vs. system or life-time) costs, which will keep the cost of the initial loan low and finance the balance through cash flow.

Energy self-reliance or independence. This, of course, became a recognized and institutionalized concern in many countries (such as Japan and the US) due to the 1973 OPEC oil embargo and the recognition of the economic power wielded by that group of producers.

International agreements. There are two types of international agreements that significantly affect or are affected by choices among energy supply options: environmental and security/defense. The most important environmental agreement is the Framework Convention on Climate Change (FCCC) signed at Rio de Janeiro during the Earth Summit in 1992, and which has since been signed by over 160 countries and ratified by over 118 countries, including the US. The overall goal of the FCCC is to stabilize emissions of greenhouse gases at a level that prevents "dangerous" interference with the climate system. As a first step, the industrialized countries agreed to voluntarily attempt to limit emissions by the year 2000 to 1990 levels. Negotiations are continuing now on a protocol for dealing with future emissions. As we will see later, this agreement tends to have a positive impact on nuclear power and renewables as energy supply options, and a strong negative impact on fossil fuels. The most important international security agreement in the context of this discussion is the Nuclear Non-proliferation Treaty (NPT) which is designed to prevent the transfer of nuclear weapons and nuclear weapons production technologies to non-nuclear weapons states. Clearly, this has had a significant negative impact on the development of civil nuclear power as a viable energy supply choice for some nations (China, for example).

Existence or absence of an established infrastructure. The existence or absence of both an electricity grid, and/or a transportation infrastructure for energy feedstocks such as coal or oil (both of which can be called the 'physical infrastructure') can have a significant impact on the choice of energy supply option. Clearly, if such a physical infrastructure does not exist it must be built and therefore financed, or an energy supply option which does not require a grid (such as some renewables, and diesel generators) will be chosen. The education or knowledge infrastructure also should impact choice: is requisite knowledge to build, maintain, and repair the chosen energy technologies available in-country or can it be easily developed?

Relationship of the choice to sustainability. 'Sustainability' is shorthand for a new paradigm of development, a way of looking at social processes that includes consideration of their impact on planetary (and therefore, local) ecology. It requires a long-term, inter-generational perspective (with significant re-definitions of and consequences for 'return on investment' concepts). It requires implementation of concepts such as 'industrial ecology' which treat industrial (and by extension, social) processes as closed systems, considering systemic consequences of all parts of the process from raw material extraction through waste treatment and disposal (Dambach 1994) where the system is a global one. It thus encompasses and attempts to reconcile concepts of both 'economic development' and environmental quality.

We have identified five primary energy supply options (fossil, nuclear, renewables, conservation, and 'unconventional') and five key factors driving choices among those options (economics, the desire for energy independence, the existence of international agreements, the existence or absence of established infrastructures, and the relationship of the choice to sustainability). We now will take the three primary supply options (fossil, nuclear, and renewables) and demonstrate in general terms how the factors driving choices among these options could play out with significant negative global consequences and that renewables can, with little additional government support, address some of the more serious of these consequences. We will conclude with an observation that renewables be considered a serious and key part of future global energy strategies.

Fossil Energy and Factors Driving Choices

As noted earlier, fossil fuels make up almost 90% of today's energy mix, and most scenarios see them occupying a similar strong position in the future. If we examine the consequences of such a mix in the light of the factors driving choices among energy supply options, we see both the logic as well as some negative consequences of such selections.

Economics. The economics for fossil fuels are generally positive. The retail cost per kwh is relatively low, averaging about 8.4 cents/kwhr in the US (EIA 1995). Fossil fuel technologies are relatively mature, which brings the cost of acquisition down (R&D costs have generally been amortized, and economies of scale and production exist). Coal is generally available worldwide, however, there is uncertainty about future prices. Natural gas prices in the US are dropping because of the profound systemic changes induced by deregulation, and are generally falling world-wide because of recently increased reserve estimates. Environmental costs are externalized and generally not included in cost per kwh calculations. In the financing arena, the mature technology reduces uncertainty and so reassures investors and makes financing relatively easy to acquire for fossil fuel-fired power plants. For distributed power applications such as diesel and small gas turbines, installation (up-front) costs are low, so, once again, financing is relatively easy to obtain.

Desire for energy independence. The primary negative for fossil fuels is, of course, in the transportation sector which is almost completely dependent upon oil and so upon the limited number of oil-producing countries in the world. In terms of electricity generation, fossil fuels, particularly coal, look attractive as coal is well distributed geographically (China has significant reserves, although it has infrastructure concerns as most of its coal reserves are located in the northern part of the country while consumption centers are in the south and southeast, and transportation is poor) and its international movement is not controlled by any cartel-like organization such as OPEC.

Existence of international agreements. Fossil fuels are a big negative in the environmental arena. Fossil fuel-fired power plants and the transportation sector (which is almost entirely based on oil) will be responsible for most CO_2 and SO_2 production, affecting compliance with the FCCC. Developing regions will significantly increase their emissions of greenhouse gasses in the near future. In the next 20 years, the relative share of CO_2 emissions from the developing world, including China, could increase from 19% to 48%. Furthermore, if developing countries achieve the level of electrification expected by 2040, their relative share would increase to over 70%, even though per capita consumption would remain far below current industrialized world levels (Drennen 1993).

A unilateral response by industrialized nations thus would be largely ineffective at reducing global greenhouse gas emissions. The logic of energy choices exercised by the developing countries will have a profound impact in this arena.

The second area of international agreements, that of security treaties related to defense concerns such as proliferation, is positive for fossil fuels. Although there are agreements through the IEA for mutual response in the event of another Arab oil embargo (a commitment that seemed to be demonstrated through the coalition assembled during the Gulf War), the movement of fossil fuels *per se* is not restricted by international agreements of this sort - in fact, it is enhanced and encouraged.

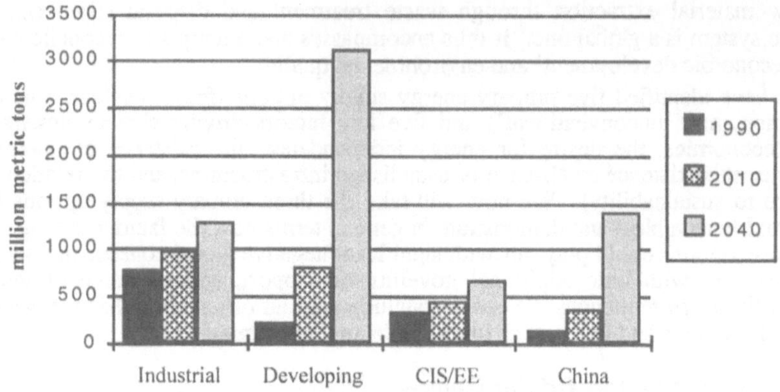

Figure 4: CO_2 **Emissions from Electricity**

Source: Drennen 1993

Existence or absence of established infrastructures. In terms of physical infrastructures, fossil fuels rack up both negative and positive points, but make strong positive marks in the education/knowledge infrastructure arena. Most developing countries do not have the grid required to distribute electricity to rural areas (and, in some cases, to urban areas) and so would have to finance and construct it. However, the maturity of the technology and construction techniques make the acquisition of such financing relatively easy - and, in those areas where grid development does not seem to be feasible, the relatively low capital costs required for diesel and small gas turbine generators favor the use of fossil fuel. It is here, too, the absence of coal transportation networks in China becomes relevant. However, again, the technology for constructing such transportation networks is well known and easily available. Finally, the knowledge or education regarding combustion technology is easily available and already relatively widespread.

Relationship to sustainability. Fossil comes out with a strong negative here. It is a depletable resource in all its forms (coal, oil, natural gas), and its consumption contributes significantly to environmental degradation through the production of CO_2 and other greenhouse gasses.

Fossil fuel as an energy supply choice provides a mixed bag of positives and negatives when evaluated against the factors driving such choices.

In summary, we find on the positive side that historic investments in the knowledge infrastructure have led fossil fuel technologies to their current position as mature technologies. This now contributes to the positive economics of fossil fuel-generated energy, including its

	Economics		Energy	Internat'l Agreements		Infrastructure		Sustainability
	Production	Financing	Independence	Environ.	Security	Phys.	Knowledge	
Fossil	+	+	+/-	-	-/+	-/+	+	-

Figure 5: Summary of Fossil Energy as an Energy Fuel Option

relatively low kwh cost and its good borrowing ability. The big negatives for fossil fuel as an energy option are in the environmental arena (particularly with CO_2 emissions and other greenhouse gases, and the way in which they will increase over time) and in sustainability (as we near the limits of recoverable reserves, there will be an adverse affect on price). Note, too, that both negatives are exacerbated by time, i.e. if we take the long-term perspective required by such concepts as industrial ecology and sustainable development, fossil energy looks like an increasingly poor energy supply option.

Nuclear Energy and Factors Driving Choices

Nuclear energy also ends up as a mixed bag. While its economics are generally good, they are negatively influenced by uncertainties, and there are significant negatives in the security and environmental/sustainability arenas.

Economics. The economics of nuclear power are complicated by the uncertainties introduced by regulatory issues. The cost of production of nuclear power has the potential to be competitive, and the fuel (uranium) is cheap. However, increasingly strict regulations which are generally designed to internalize health and environmental costs, and the regulatory uncertainties present both in the US and in other countries, drive up costs. These same uncertainties make the cost of plant construction variable and so difficult to finance through international lending institutions. However, in some cases, state-sponsored nuclear power developers (such as Electricité de France) can obtain 'self-financing,' so international lending community does not need to be involved, and nuclear power becomes a more attractive option for developing countries.

Desire for energy independence. Many countries such as Japan and France have seen nuclear energy as a path to energy independence and have explicitly designed energy strategies around it. There is a negative encountered by some developing countries as they attempt to follow this path in the international controls put on the movement of nuclear technologies and special nuclear materials in the name of non-proliferation.

Existence of international agreements. The primary concern in the environmental arena is the disposal of waste and the potential for generating radioactive plumes in the event of an accident. (For example, Indonesia's announced plans for constructing seven to twelve nuclear power plants in the 600-1000 MW range each caused international concern because of the archipelago's location on the geologically unstable Pacific 'Ring of Fire.') (Habibie 1994). However, the production of power with nuclear technology does not generate greenhouse gasses; hence it can be argued that use of nuclear power clearly supports the FCCC. It is under international security agreements, particularly the NPT, that nuclear power hits its biggest negative. In the interest of limiting the spread of nuclear weapons, the international movement of all nuclear technologies and materials is strictly controlled and may be denied to nations with declared intentions of using it only for peaceful purposes if the international community deems otherwise (the recent discussion of the sale of Russian technology and equipment to Iran is a case in point).

Existence or absence of established infrastructures. Nuclear power requires a grid; there are no distributed power options. Hence, nuclear power as an energy supply option encounters the same concerns as fossil in this context with no mitigating technologies such as diesel generators or small gas turbines. In terms of knowledge infrastructures, while knowledge of nuclear power technologies is fairly well developed, its dissemination is controlled due to (among other reasons) non-proliferation concerns. As a result, there also is an element of 'technology fear' connected with nuclear energy that may not be present with fossil energy knowledge.

Relationship of the choice to sustainability. Nuclear energy has the potential to show favorable marks in this column if the waste problem can be solved in a fashion that meets environmental concerns but does not threaten international security as does (some believe) current reprocessing techniques, and if the public can accept a level of risk at least comparable with if not better than fossil fuel-fired plants. As the state of the art currently stands however, nuclear energy must be given a negative here.

The big negatives for nuclear energy are in the international security and environment/sustainability columns, both of which have led to increased regulatory requirements which have had a negative impact on economics. Also unfavorable for developing countries is the requirement for a grid. Positives are in nuclear energy's contribution to energy independence and the leveraging of an existing knowledge infrastructure which has had a positive impact on economics.

	Economics		Energy	Internat'l Agreements		Infrastructure		Sustainability
	Production	Financing	Independence	Environ.	Security	Phys.	Knowledge	
Fossil	+	+	+/-	-	-/+	-/+	+	-
Nuclear	+/-	+	+/-	+/-	-	-	+/-	+/-

Figure 6: Summary of Nuclear Energy as an Energy Fuel Option

We still don't have an energy option which is attractive in the context of the emerging consumers (the developing nations, including China and India) which will see significant increases in the amount of energy consumed particularly in rural areas, generally do not have a grid structure in place to service these new consumers, are seeking energy independence to avoid slipping into dependencies of 'neo-colonialism,' and which may be politically unstable enough to pose potential proliferation concerns. Finally, the burgeoning world population is stimulating serious concerns about global sustainability, which neither fossil nor nuclear energy addresses favorably.

Renewable Energy and Factors Driving Choices

Rounding out our picture of available energy supply options, we now turn to a consideration of renewables. The analysis shows that, while negative in some columns such as cost of production, renewables do show positives in areas missed by fossil and nuclear, such as sustainability. This raises the suggestion of a *portfolio* of options, and suggests that a conscious, deliberate, and well-considered global energy strategy might require investment by both government and private sector in different parts of the energy supply mix.

Economics. The economics of non-hydro renewables generally are unfavorable, particularly when compared with other energy supply options.[1] The costs of production generally are high, although they have decreased significantly over the last 15 years and are approaching comparability with fossil fuel-generated electricity.[2]

However, if environmental costs, currently treated as externalities in fossil and nuclear fuel costs, were internalized, the costs of renewable-generated energy would be much more competitive. Furthermore, renewables are very competitive with distributed power options such as diesel generators in certain niche but important off-grid applications. Finally, financing is often difficult to obtain for renewable energy projects,

[1] The economic viability of non-hydro renewables has recently been challenged in a report by the Center for Energy and Economic Development that concluded that renewable energy will be unable to contribute significantly to the US's electric power needs in the next 15 years without massive federal subsidies. The report was strongly rebutted by the US Department of Energy in a reponse prepared by the National Renewable Energy Laboratory (cf The Energy Daily 1995).

[2] In the discussion referred to above between the US Department of Energy and the Center for Energy and Economic Development, the Center claimed that cost per kwh of renewables would level off at today's cost; the Department of Energy argued for a continued decline (see Figure 7).

because technologies are immature and unproven, the up-front installation costs can be higher than alternatives, although system life costs may be competitive because there are no add-on fuel costs, and because of certain environmental considerations, particularly those caused by hydropower's large footprint.[3]

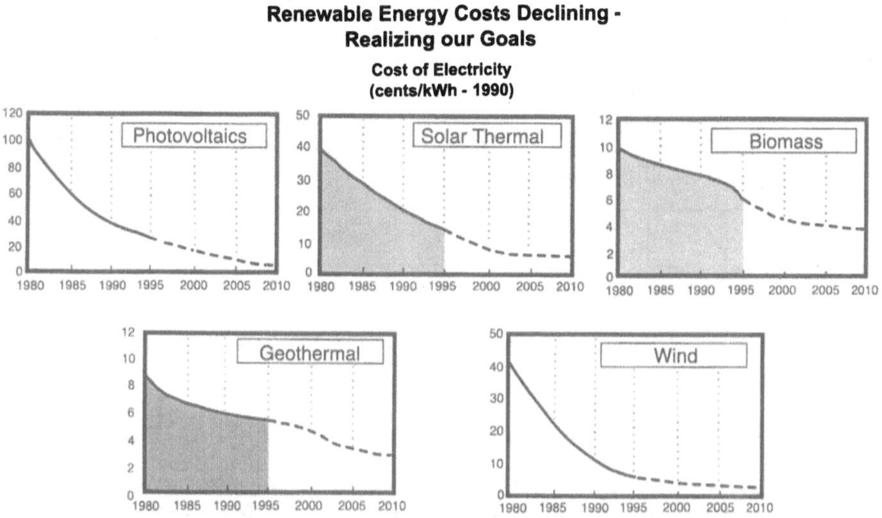

Figure 7: Cost of electricity for renewable options

source: US Department of Energy, Energy Efficiency and Renewable Energy 1995

Energy independence. Renewables place high here, as every country can do some sort and some amount of renewable energy production.

International agreements. Production of energy with renewable sources such as water or the sun does not generate greenhouse gasses; hence renewables comes up positive against the FCCC. However, there are some serious environmental concerns associated with some renewable deployments, such as the Three Gorges Dam in China - in this case, concerns associated both with the large footprint required and the destruction of habitat and changes in water flows occasioned by the dam. Renewables raise no concerns regarding international security.

Existence or absence of an established infrastructure. Here, again, renewables come out both positive and negative. There are many renewable technologies that are highly suitable for distributed applications, and hence to not require development of a grid. The nature of the feedstock also means that there are no transportation issues with which to deal. However, the installation, repair, and maintenance knowledge infrastructures are in their infancy, perhaps raising concerns of intellectual dependency (neo-colonialism) in some developing countries.

Relationship of the choice to sustainability. It is here that renewable energy options shine. Properly implemented, they will act as effective closed systems, generating little or no waste and consuming little or no irreplaceable resources.

Although they are improving, the economics of renewables are still unfavorable, particularly when compared with other energy options. Costs per kwh are high, although they are rapidly decreasing and would look much better compared to other options if environmental externalities were considered in all energy pricing. The up-front cost of

[3] The Clinton administration recently announced that it opposes credit from the US Export-Import Bank for companies seeking to participate in China's Three Gorges Dam project. Opposition was due to "environmental concerns," according to White House press secretary, Mike McCurry (*The China Energy Report, 1995*).

system installation can be high, making renewables an unattractive option for financing systems that require a near-term return on investment and so favor smaller loans as well as loans directed at grid-connected applications which can generate a much higher cash flow than can off-grid connections which generally target rural consumers. However, for this same reason (grid independence) renewables speaks well to developing countries without established infrastructures, and also supports energy independence, although the absence of the knowledge infrastructure does raise fears of dependence. Renewables also show positives in key columns which speak to the needs of the global community - sustainability and international agreements.

	Economics		Energy	Internat'l Agreements		Infrastructure		Sustainability
	Production	Financing	Independence	Environ.	Security	Phys.	Knowledge	
Fossil	+	+	+/-	-	-/+	-/+	+	-
Nuclear	+/-	+	+/-	+/-	-	-	+/-	+/-
Renewables	-	-	+/-	+	+	+	-	+

Figure 8: Summary of Renewable Energy as an Energy Fuel Option

IMPLICATIONS OF ENERGY CHOICES FOR GLOBAL QUALITY OF LIFE

The geopolitical distribution of the bulk of new consumers of energy - in the developing countries, most particularly China and India - say that we all will have to do something about such long-term and systems problems as planetary sustainability. As we showed earlier with the projections on CO_2 production, this must indeed be a joint effort: the industrialized nations cannot do it alone. However, given that the factors influencing choices among energy supply options generally are manifest in ways that favor industrialized nations (the most clear example being that of financing, where the system has made it most difficult to obtain financing for that energy application where developing countries need it most - rural electrification), it is most likely that our future energy supply mix will look very similar to that which we have today. I would like to make a few suggestions that may alter that mix, and then leave you with some questions.

How to Improve the End Game

If our analysis is reasonably accurate, it would suggest that the best global energy supply mix would be just that - a mix or portfolio of energy options, providing affordable energy when and where needed, offering a hedge against unforeseen geopolitical events that might radically alter the flow of energy feedstocks around the globe, protecting against the acquisition of nuclear weapons knowledge by 'bad actors,' and doing all this in a fashion that will allow planetary longevity. As our global mix currently stands, renewables are the weak element in our portfolio. However, with a judicious mix of government and industry investment, we can significantly strengthen the offering. Government can play an important role in financing the R&D that is needed to make the technologies market-attractive, at which point the private sector should pick up the ball. A relatively small investment in education and outreach programs, again on the part of the government, will develop the knowledge infrastructure required to disseminate renewables into remote areas of developing countries, areas to which they are eminently suited. Finally, we collectively need to change our paradigms, to think long-term, inter-generationally, in terms of sustainability. Such a paradigm shift would have natural fall-out in the global financing system, internalizing what are now 'environmental externalities' and suggesting a life-time view of energy systems that would incorporate fuel costs, particularly for fossil fuel-based energy production systems. If we can do these things, we can, for a relatively small social cost, develop an 'energy insurance portfolio' that would help improve the global quality of life without compromising key social goals. And a viable renewable sector is key to that portfolio.

Now for the questions. The US has very effectively build a defense industrial base to assure our national ability to respond to aggression and to defend us against any other country's attempt to detract from our life, liberty, or pursuit of happiness. We should see the energy industrial base in the US as the equivalent in its importance to preserve these rights for Americans and to respond to the global economic market that is going to develop in the next two decades and that should spread some of these opportunities to others. The US has made a number of (aborted) attempts to take this high ground, but it has regularly failed. What is the root cause of these failures? How do we deal with them? Are they all economic or are they political and competitive? Are other countries responding to this market and moral message and leaving us behind?

We all - whether we live in industrialized or developing countries - have a common investment in the future of our planet. Renewable energy as a significant part of our global energy portfolio can help make a real contribution to sustainability. The energy sources will be continuously available: exploitation of them is limited only by our imagination. We must see beyond our current paradigms, move beyond inter- national competition to a concerted global effort, and use the closed system concepts of industrial ecology to begin development of a reasonable and considered response to this problem. Should we fail to assume this responsibility, we will consign it to our children and grandchildren who will have to address it in less time and with fewer resources.

ACKNOWLEDGMENTS

The author would like to thank the following individuals for their help in developing this paper: Dan Arvizu, Tom Drennen, Dennis Engi, and Jessica Glicken, all of Sandia National Laboratories.

REFERENCES

Dambach, B. F., 1994, *Translating Industrial Ecology Into Design for Environment.*

Drennen, T., 1993, *Economic Development and Climate Change: Analyzing the International Response*, Ph.D. dissertation, Cornell University.

Habibie, B. J., 1994, "The Development of Nuclear Power Plants in Indonesia", Agency for Assessment and Application of Technology, January 1994.

Lu, Y., 1993, *Fueling one billion: an insider's story of Chinese energy policy development*, The Washington Institute Press, Washington, DC.

Mellecker, D., 1995, "Turning the Corner" *Solar Industry Journal* Second Quarter 1995: 22-26.

China Energy Report, The, 1995, Volume 2, Number 10, October 1995 (published by Strategic Marketing Inc., Hartsdale, NY).

Energy Daily, The, 1995, Volume 23, Number 214, November 13, 1995.

Economist, The, 1994, "Energy Survey" June 18, 1994 (special supplement, 18 pp.).

US Department of Energy, Energy Information Agency, 1995, International Energy Outlook.

EIA "Monthly Energy Review October 1995".

1994 Annual Energy Review.

US Department of Energy, Office of Energy Efficiency & Renewable Energy 1995, Office of Utility Technologies.

1993 EIA, International Energy Annual, Table 2.9.

INDEX